Corrosion monitoring in nuclear systems: research and applications

European Federation of Corrosion Publications
NUMBER 56

Corrosion monitoring in nuclear systems: research and applications

Edited by

Stefan Ritter & Anders Molander

**Published for the European Federation of Corrosion
by CRC Press
on behalf of
The Institute of Materials, Minerals & Mining**

CRC Press
Taylor & Francis Group
Boca Raton London New York

CRC Press is an imprint of the
Taylor & Francis Group, an **informa** business

The Institute of Materials, Minerals & Mining

First published 2010 by Maney Publishing

Published 2019 by CRC Press
Taylor & Francis Group
6000 Broken Sound Parkway NW, Suite 300
Boca Raton, FL 33487-2742

First issued in paperback 2019

No claim to original U.S. Government works

ISBN 13: 978-0-367-44597-3 (pbk)
ISBN 13: 978-1-906540-98-2 (hbk)
ISSN 1354-5116

Visit the Taylor & Francis Web site at
http://www.taylorandfrancis.com

and the CRC Press Web site at
http://www.crcpress.com

Cover photo by Yaw-Ming Chen, MCL/ITRI, Taiwan (from Chapter 12: pipe-thinning monitor; potential drop measurements on a carbon steel elbow pipe).

Contents

Series introduction   *vii*

Volumes in the EFC series   *ix*

Preface   *xiv*

Section 1: Introduction
1.   Corrosion monitoring – what's the point?
        Bob Cottis   1

Section 2: Research on Corrosion Monitoring for Nuclear Systems
2.   Introduction to different electrochemical corrosion
        monitoring techniques
        Rik-Wouter Bosch & Walter F. Bogaerts   5

3.   Electrochemical noise generated during stress corrosion
        cracking in high-temperature water and its relationship to
        repassivation kinetics
        *Rik-Wouter Bosch, Marc Vankeerberghen, Serguei Gavrilov
        & Steven Van Dyck*   32

4.   Detection of stress corrosion cracking in a simulated BWR
        environment by combined electrochemical potential noise and
        direct current potential drop measurements
        Stefan Ritter & Hans-Peter Seifert   46

5.   Electrochemical noise measurements of stainless steel in
        high-temperature water
        Carlos R. Arganis-Juarez, José M. Malo & J. Uruchurtu   63

6.   Stochastic approach to electrochemical noise analysis of SCC
        of Alloy 600 SG tube in caustic environments at high temperature
        *Sung-Woo Kim, Hong-Pyo Kim, Seong-Sik Hwang, Dong-Jin Kim,
        Joung-Soo Kim, Yun-Soo Lim, Sung-Soo Kim & Man-Kyo Jung*   81

7.   In-situ electrochemical impedance and noise measurements of
        corroding stainless steel in high-temperature water
        Jan Macák, Petr Sajdl, Pavel Kučera, Radek Novotný & Jan Vošta 96

8.   Monitoring of stress corrosion cracking of sensitised 304H
        stainless steel by electrochemical methods and acoustic emission
        Wenzhong Zhang, Lucia Dunbar & David Tice   120

9.   Corrosion monitoring of carbon steel in pasty clayey mixture
as a function of temperature
Christian Bataillon, Frantz A. Martin & Marc Roy   143

***Section 3: Corrosion Monitoring Applications in Nuclear Power Plants
(Light Water Reactors)***

10.   Corrosion monitoring applications in nuclear power
plants – a review
Shunsuke Uchida   158

11.   In-core corrosion monitoring in the Halden test reactor
Peter Bennett & Torill Karlsen   171

12.   Corrosion monitoring applications in Taiwan's nuclear
power plants
Hsuan-Chin Lai, Wei-Yun Mao & Charles Fang Chu   194

13.   A new technique for in-plant high mass transfer rate
ECP monitoring
Anders Molander & Mats Ullberg   211

14.   Precipitation and transformation of iron species in the
presence of oxygen and hydrazine in a simulated stainless steel
feed water system
Jerzy A. Sawicki, Anders Molander & Agnès Stutzmann   221

15.   An electrochemical sensor array for in-situ measurements of
hydrogen peroxide concentration in high-temperature water
Shunsuke Uchida, Tomonori Satoh, Yoshiyuki Satoh & Yoichi Wada   239

Index   255

European Federation of Corrosion (EFC) publications: Series introduction

The European Federation of Corrosion (EFC), incorporated in Belgium, was founded in 1955 with the purpose of promoting European cooperation in the fields of research into corrosion and corrosion prevention.

Membership of the EFC is based upon participation by corrosion societies and committees in technical Working Parties. Member societies appoint delegates to Working Parties, whose membership is expanded by personal corresponding membership.

The activities of the Working Parties cover corrosion topics associated with inhibition, education, reinforcement in concrete, microbial effects, hot gases and combustion products, environment-sensitive fracture, marine environments, drinking water systems, refineries, surface science, physico-chemical methods of measurement, the nuclear industry, the automotive industry, coatings, polymeric materials, tribo-corrosion, the oil and gas industry and the archaeological heritage. Working Parties and Task Forces on other topics are established as required.

The Working Parties function in various ways, e.g. by preparing reports, organising symposia, conducting intensive courses and producing instructional material, including films. The activities of Working Parties are coordinated, through a Science and Technology Advisory Committee, by the Scientific Secretary. The administration of the EFC is handled by three Secretariats: DECHEMA e.V. in Germany, the Fédération Française pour les sciences de la Chimie (formely Société de Chimie Industrielle) in France, and The Institute of Materials, Minerals and Mining in the UK. These three Secretariats meet at the Board of Administrators of the EFC. There is an annual General Assembly at which delegates from all member societies meet to determine and approve EFC policy. News of EFC activities, forthcoming conferences, courses, etc., is published in a range of accredited corrosion and certain other journals throughout Europe. More detailed descriptions of activities are given in a Newsletter prepared by the Scientific Secretary.

The output of the EFC takes various forms. Papers on particular topics, e.g. reviews or results of experimental work, may be published in scientific and technical journals in one or more countries in Europe. Conference proceedings are often published by the organisation responsible for the conference.

In 1987 the, then, Institute of Metals was appointed as the official EFC publisher. Although the arrangement is non-exclusive and other routes for publication are still available, it is expected that the Working Parties of the EFC will use The Institute of Materials, Minerals and Mining for publication of reports, proceedings, etc., wherever possible.

The name of The Institute of Metals was changed to The Institute of Materials (IoM) on 1 January 1992 and to The Institute of Materials, Minerals and Mining with effect from 26 June 2002. The series is now published by Maney Publishing on behalf of The Institute of Materials, Minerals and Mining.

P. McIntyre
EFC Series Editor
The Institute of Materials, Minerals and Mining, London, UK

EFC Secretariats are located at:

Dr B. A. Rickinson
European Federation of Corrosion, The Institute of Materials, Minerals and Mining, 1 Carlton House Terrace, London SW1Y 5AF, UK

Mr. M. Roche
Fédération Européenne de la Corrosion, Fédération Française pour les sciences de la Chimie, 28 rue Saint-Dominique, F-75007 Paris, France

Dr. W. Meier
Europäische Föderation Korrosion, DECHEMA e.V., Theodor-Heuss-Allee 25, D-60486 Frankfurt-am-Main, Germany

Volumes in the EFC series

* indicates volume out of print

1 **Corrosion in the nuclear industry**
Prepared by Working Party 4 on Nuclear Corrosion*

2 **Practical corrosion principles**
Prepared by Working Party 7 on Corrosion Education*

3 **General guidelines for corrosion testing of materials for marine applications**
Prepared by Working Party 9 on Marine Corrosion*

4 **Guidelines on electrochemical corrosion measurements**
Prepared by Working Party 8 on Physico-Chemical Methods of Corrosion Testing

5 **Illustrated case histories of marine corrosion**
Prepared by Working Party 9 on Marine Corrosion

6 **Corrosion education manual**
Prepared by Working Party 7 on Corrosion Education

7 **Corrosion problems related to nuclear waste disposal**
Prepared by Working Party 4 on Nuclear Corrosion

8 **Microbial corrosion**
Prepared by Working Party 10 on Microbial Corrosion*

9 **Microbiological degradation of materials and methods of protection**
Prepared by Working Party 10 on Microbial Corrosion

10 **Marine corrosion of stainless steels: chlorination and microbial effects**
Prepared by Working Party 9 on Marine Corrosion

11 **Corrosion inhibitors**
Prepared by the Working Party on Inhibitors*

12 **Modifications of passive films**
Prepared by Working Party 6 on Surface Science*

13 **Predicting CO_2 corrosion in the oil and gas industry**
Prepared by Working Party 13 on Corrosion in Oil and Gas Production*

14 **Guidelines for methods of testing and research in high temperature corrosion**
Prepared by Working Party 3 on Corrosion by Hot Gases and Combustion Products

15   **Microbial corrosion: Proceedings of the 3rd International EFC Workshop**
      Prepared by Working Party 10 on Microbial Corrosion

16   **Guidelines on materials requirements for carbon and low alloy steels for H₂S-containing environments in oil and gas production (3rd Edition)**
      Prepared by Working Party 13 on Corrosion in Oil and Gas Production

17   **Corrosion resistant alloys for oil and gas production: guidance on general requirements and test methods for H₂S service**
      Prepared by Working Party 13 on Corrosion in Oil and Gas Production

18   **Stainless steel in concrete: state of the art report**
      Prepared by Working Party 11 on Corrosion of Steel in Concrete

19   **Sea water corrosion of stainless steels: mechanisms and experiences**
      Prepared by Working Party 9 on Marine Corrosion and Working Party 10 on Microbial Corrosion

20   **Organic and inorganic coatings for corrosion prevention: research and experiences**
      Papers from EUROCORR '96

21   **Corrosion-deformation interactions**
      CDI '96 in conjunction with EUROCORR '96

22   **Aspects of microbially induced corrosion**
      Papers from EUROCORR '96 and EFC Working Party 10 on Microbial Corrosion

23   **CO₂ corrosion control in oil and gas production: design considerations**
      Prepared by Working Party 13 on Corrosion in Oil and Gas Production

24   **Electrochemical rehabilitation methods for reinforced concrete structures: a state of the art report**
      Prepared by Working Party 11 on Corrosion of Steel in Concrete

25   **Corrosion of reinforcement in concrete: monitoring, prevention and rehabilitation**
      Papers from EUROCORR '97

26   **Advances in corrosion control and materials in oil and gas production**
      Papers from EUROCORR '97 and EUROCORR '98

27   **Cyclic oxidation of high temperature materials**
      Proceedings of an EFC Workshop, Frankfurt/Main, 1999

28   **Electrochemical approach to selected corrosion and corrosion control**
      Papers from the 50th ISE Meeting, Pavia, 1999

29   **Microbial corrosion: proceedings of the 4th International EFC Workshop**
      Prepared by the Working Party on Microbial Corrosion

30   **Survey of literature on crevice corrosion (1979–1998): mechanisms, test methods and results, practical experience, protective measures and monitoring**
      Prepared by F. P. Ijsseling and Working Party 9 on Marine Corrosion

31   **Corrosion of reinforcement in concrete: corrosion mechanisms and corrosion protection**
Papers from EUROCORR '99 and Working Party 11 on Corrosion of Steel in Concrete

32   **Guidelines for the compilation of corrosion cost data and for the calculation of the life cycle cost of corrosion: a working party report**
Prepared by Working Party 13 on Corrosion in Oil and Gas Production

33   **Marine corrosion of stainless steels: testing, selection, experience, protection and monitoring**
Edited by D. Féron on behalf of Working Party 9 on Marine Corrosion

34   **Lifetime modelling of high temperature corrosion processes**
Proceedings of an EFC Workshop 2001 Edited by M. Schütze, W. J. Quadakkers and J. R. Nicholls

35   **Corrosion inhibitors for steel in concrete**
Prepared by B. Elsener with support from a Task Group of Working Party 11 on Corrosion of Steel in Concrete

36   **Prediction of long term corrosion behaviour in nuclear waste systems**
Edited by D. Féron on behalf of Working Party 4 on Nuclear Corrosion

37   **Test methods for assessing the susceptibility of prestressing steels to hydrogen induced stress corrosion cracking**
By B. Isecke on behalf of Working Party 11 on Corrosion of Steel in Concrete

38   **Corrosion of reinforcement in concrete: mechanisms, monitoring, inhibitors and rehabilitation techniques**
Edited by M. Raupach, B. Elsener, R. Polder and J. Mietz on behalf of Working Party 11 on Corrosion of Steel in Concrete

39   **The use of corrosion inhibitors in oil and gas production**
Edited by J. W. Palmer, W. Hedges and J. L. Dawson on behalf of Working Party 13 on Corrosion in Oil and Gas Production

40   **Control of corrosion in cooling waters**
Edited by J. D. Harston and F. Ropital on behalf of Working Party 15 on Corrosion in the Refinery Industry

41   **Metal dusting, carburisation and nitridation**
Edited by H. Grabke and M. Schütze on behalf of Working Party 3 on Corrosion by Hot Gases and Combustion Products

42   **Corrosion in refineries**
Edited by J. D. Harston and F. Ropital on behalf of Working Party 15 on Corrosion in the Refinery Industry

43   **The electrochemistry and characteristics of embeddable reference electrodes for concrete**
Prepared by R. Myrdal on behalf of Working Party 11 on Corrosion of Steel in Concrete

44   **The use of electrochemical scanning tunnelling microscopy (EC-STM) in corrosion analysis: reference material and procedural guidelines**
Prepared by R. Lindström, V. Maurice, L. Klein and P. Marcus on behalf of Working Party 6 on Surface Science

45   **Local probe techniques for corrosion research**
Edited by R. Oltra on behalf of Working Party 8 on Physico-Chemical Methods of Corrosion Testing

46   **Amine unit corrosion survey**
Edited by J. D. Harston and F. Ropital on behalf of Working Party 15 on Corrosion in the Refinery Industry

47   **Novel approaches to the improvement of high temperature corrosion resistance**
Edited by M. Schütze and W. Quadakkers on behalf of Working Party 3 on Corrosion by Hot Gases and Combustion Products

48   **Corrosion of metallic heritage artefacts: investigation, conservation and prediction of long term behaviour**
Edited by P. Dillmann, G. Béranger, P. Piccardo and H. Matthiesen on behalf of Working Party 4 on Nuclear Corrosion

49   **Electrochemistry in light water reactors: reference electrodes, measurement, corrosion and tribocorrosion**
Edited by R.-W. Bosch, D. Féron and J.-P. Celis on behalf of Working Party 4 on Nuclear Corrosion

50   **Corrosion behaviour and protection of copper and aluminium alloys in seawater**
Edited by D. Féron on behalf of Working Party 9 on Marine Corrosion

51   **Corrosion issues in light water reactors: stress corrosion cracking**
Edited by D. Féron and J-M. Olive on behalf of Working Party 4 on Nuclear Corrosion

52   **Progress in Corrosion – The first 50 years of the EFC**
Edited by P. McIntyre and J. Vogelsang on behalf of the EFC Science and Technology Advisory Committee

53   **Standardisation of thermal cycling exposure testing**
Edited by M. Schütze and M. Malessa on behalf of Working Party 3 on Corrosion by Hot Gases and Combustion Products

54   **Innovative pre-treatment techniques to prevent corrosion of metallic surfaces**
Edited by L. Fedrizzi, H. Terryn and A. Simões on behalf of Working Party 14 on Coatings

55   **Corrosion-under-insulation (CUI) guidelines**
Prepared by S. Winnik on behalf of Working Party 13 on Corrosion in Oil and Gas Production and Working Party 15 on Corrosion in the Refinery Industry

56   **Corrosion monitoring in nuclear systems: research and applications**
Edited by S. Ritter and A. Molander

58   **Self-healing properties of new surface treatments**
Edited by L. Fedrizzi, W. Fürbeth and F. Montemor

60   **Methodology of crevice corrosion testing for stainless steels in natural and treated seawaters**
Edited by U. Kivisäkk, B. Espelid and D. Féron

All volumes are available from Maney Publishing or its North American distributor. See http://maney.co.uk/index.php/series/efc_series/

Preface

Even though the plant availability of nuclear power plants has increased and radiation build-up has been effectively mitigated, corrosion continues to influence plant performance and availability. Recent examples are stress corrosion cracking (SCC) attacks on type 316L stainless steel in boiling water reactors (BWRs), attacks on nickel-base alloys in pressurised water reactors (PWRs) and irradiation-assisted SCC incidents in both types of reactor. Therefore, reliable prediction of the corrosion behaviour of given materials is considered important for the future operation of existing plants, which are commonly subject to life-time extension and power uprating. In addition, managing corrosion problems in future nuclear waste disposal systems is another task where reliable corrosion monitoring techniques could help to further increase safety. Tremendous effort has been spent during the years to mitigate corrosion and improve the water chemistry of nuclear power reactors. However, actual in-plant corrosion monitoring has been performed or developed for only a very limited number of applications. Therefore a workshop was arranged by Working Party 4 ('Nuclear Corrosion') of the European Federation of Corrosion (EFC) and the European Cooperative Group on Corrosion Monitoring of Nuclear Materials (ECG-COMON, www.ecg-comon.org) during EUROCORR 2007, which aimed to highlight such work and identify monitoring techniques with potential for future application. Contributions were invited covering research on or the application of corrosion monitoring techniques (e.g., electrochemical noise, direct and alternate current potential drop, acoustic emission, electrochemical impedance spectroscopy, monitoring corrosion potentials, activity monitoring, etc.) for all types of fission reactors (including the secondary side) and nuclear waste disposal systems. Some of the responses obtained, including additional contributions, are presented here. The book is divided into two main parts: 'Research on corrosion monitoring for nuclear systems' and 'Corrosion monitoring applications in nuclear power plants (light water reactors)'.

The first part deals mainly with detection of SCC initiation in nuclear power plant environments – essentially high-temperature water at around 300°C. Research work on SCC for nuclear applications has, in the past decades, focused more on crack growth. A reliable well-known technique, the potential drop technique, is used for crack growth measurements and very low growth rates can be accurately monitored. A large database has been created for various low-alloy reactor pressure vessel steels, stainless steels and nickel-base alloys in BWR and PWR environments. The effects of water chemistry have been studied, including the presence of various impurities and additives, as well as the effects of material properties, such as the presence of cold work. It was generally found that all materials can suffer from cracking in tests with pre-cracked specimens and where initiation of SCC was enhanced by dynamic loading conditions. Through such tests, the available knowledge of SCC conditions has been significantly improved. However, there are also examples of laboratory cracking

xiv

of materials known to be very resistant in plants and the observed crack growth rates were occasionally significantly higher than expected from plant experience. Such contradictions and differences might be one reason for the increased interest in initiation studies and initiation data. Unfortunately, for initiation tests, there are no verified methods analogous to the potential drop technique for crack growth investigations. Initiation tests are often performed using passively loaded specimens such as U-bends or C-rings, which are exposed to simulated plant environments and periodically inspected for cracks, or small specimens are used, where a complete failure of the specimen indicates SCC initiation. The test time can be significant – periods of years for highly resistant materials – and some materials do not crack at all in the common types of initiation tests. Material (including stress) and environmental conditions must be adjusted more aggressively to obtain cracking within a reasonable time and a proper experimental set-up with an in-situ SCC initiation monitoring technique is necessary for future initiation investigations. The formation of the ECG-COMON is an attempt to improve and develop methods for initiation studies by gathering a critical mass of international experts in this field. Various methods, but especially the electrochemical noise technique, are being investigated in this group as promising methods to detect the early stages of SCC initiation (or other types of localised corrosion). The chapters in this section demonstrate the state-of-the-art in this area of research. Very promising results are being obtained but some limitations of the monitoring methods are also becoming obvious. But it is also shown that further work is needed before, for example, electrochemical noise can be established as a reliable monitoring technique for SCC initiation in high-temperature water environments. Electrochemical techniques have also been tested in other nuclear systems, e.g. nuclear waste storage systems of which one important example is presented in this book.

In the second part of the book, work with the goal of developing in-situ techniques and some application examples of monitoring methods are presented. The contributions in this book emphasise the importance of corrosion monitoring to limit corrosion in power plants and other nuclear applications. The improvement of existing techniques and the progress of new developments are pointed out. Electrochemical noise measurements have not been applied for real in-plant measurements so far and it will take time before the technique is sufficiently developed for such applications. However, there are other techniques which have already been employed in plants for many years. These techniques are usually less informative than, for example, noise measurements could be, but are still extremely valuable. The most commonly used technique is monitoring of corrosion potentials (often called ECP). The measurement of ECP is one of the very few techniques that can be applied directly within the system studied. In this way, sampling, which can be very difficult or even impossible at high temperature and pressure, can be avoided.

To further improve corrosion monitoring techniques and to limit corrosion and activity build-up is a future challenge for both researchers and plant chemists. Considering the present nuclear revival and the increased public confidence in nuclear power in several countries, monitoring techniques, together with optimised materials behaviour and water chemistry conditions, will be particularly important in maintaining stable operating conditions in power plants and safe long-term storage conditions for radioactive waste.

We hope this book will prove useful for scientists and engineers in future corrosion mitigation work. Finally, we would like to thank all of the contributors to the workshop and to this book as well as the reviewers. We also appreciate the support from EFC Working Party 4 and ECG-COMON. Without their participation and enthusiasm, neither the workshop nor this book would have been possible.

Stefan Ritter, Anders Molander and Damien Féron

Special thanks are expressed to the reviewers of the articles:
Christian Bataillon (CEA Saclay, France)
Thomas Dorsch (Areva NP GmbH, Germany)
Lucia C. Dunbar (Serco Assurance, UK)
Damien Féron (CEA Saclay, France)
John Hickling (CMC, Cyprus)
Petri Kinnunen (VTT, Finland)
Bruno Kursten (SCK-CEN, Belgium)
Jan Kysela (NRI, Czech Republic)
Andraz Legat (ZAG, Slovenia)
Anders Molander (Studsvik Nuclear AB, Sweden)
Stefan Ritter (PSI, Switzerland)
Timo Saario (VTT, Finland)
Sannakaisa Virtanen (University of Erlangen-Nuremberg, Germany)

Most important abbreviations

AE	Acoustic emission
BWR	Boiling water reactor
CE	Counter electrode
CERT	Constant extension rate test
CGR	Crack growth rate
CPE	Constant phase element
CS	Carbon steel
C(T)	Compact tension (specimen)
DCPD	Direct current potential drop (technique)
DH	Dissolved hydrogen
DO	Dissolved oxygen
ECN	Electrochemical current noise
ECP	Electrochemical (corrosion) potential
EFM	Electrochemical frequency modulation (= intermodulation distortion)
EIS	Electrochemical impedance spectroscopy
EN	Electrochemical noise
EPN	Electrochemical potential noise
FDCI	Frequency dependent complex impedance
HA	Harmonic analysis
$[H_2]$	Hydrogen concentration
$[H_2O_2]$	Hydrogen peroxide concentration
HWC	Hydrogen water chemistry
IGSCC	Intergranular stress corrosion cracking
LPR	Linear polarisation resistance
LVDT	Linear voltage differential transformer
LWR	Light water reactor
NWC	Normal water chemistry
$[O_2]$	Oxygen concentration
PWR	Pressurised water reactor
RE	Reference electrode
RPV	Reactor pressure vessel
SCC	Stress corrosion cracking
SEM	Scanning electron microscope
SG	Steam generator
SHE	Standard hydrogen electrode
SS	Stainless steel
SSRT	Slow strain rate test
TGSCC	Transgranular stress corrosion cracking
WE	Working electrode
YS	Yield stress

1

Corrosion monitoring – what's the point?

Bob Cottis

University of Manchester, School of Materials, P.O. Box 88,
Sackville Street, Manchester M60 1QD, UK

bob.cottis@manchester.ac.uk

1.1   Introduction

The paper presents a series of largely philosophical musings on the value, practise and processes of corrosion monitoring.

When I teach corrosion monitoring, I start by asking the students four questions:

- Does corrosion monitoring reduce the rate of corrosion?
- Does corrosion monitoring reduce the cost of corrosion?
- Does corrosion monitoring reduce the hazards of corrosion?
- Does corrosion monitoring reduce the environmental impact of corrosion?

The response of the students varies a little from year to year, but in general, the majority answer 'yes' to these questions. You may wish to consider your own answers – so that you are not tempted to read the answers before doing this, this section is continued at the end of this article.

1.2   Selection of monitoring method

The ideal corrosion monitoring method would give an accurate, real-time, online indication of many properties of the system, and it would thus have all of the following capabilities:

- Instantaneous, online measure of corrosion rate
- Instantaneous, online measure of localised corrosion type(s), location and penetration rate
- Online measure of integrated uniform metal loss (i.e. measure of the remaining thickness)
- Online measure of total local corrosion penetration (i.e. pit depth)
- Location of points of large corrosion penetration
- Detection, location and sizing of SCC and similar defects
- Measure behaviour of plant (not a probe).

Of course, there is currently no such method, and corrosion monitoring necessarily involves a series of compromises. First, we will consider the compromises involved in the use of probes.

1.3   Probes for corrosion monitoring

Many corrosion monitoring techniques (e.g. all electrochemical and weight loss methods and most electrical resistance methods) depend on the use of probes (or test

1

coupons, for which very similar considerations apply), rather than making measurements of the actual material of the plant. This is convenient for several reasons, including the easy replacement of probes, the ability to test alternative materials and the fact that many of the simpler monitoring methods require the use of a probe. However, it also introduces a number of problems:

- The material of the probe is generally different from that of the plant. While it may be possible to make probes out of the same batch of material as the plant, it is much more commonly the case that it is from a different batch and probably supplied in a different form (e.g. rod rather than plate). Thus it is likely to have a different chemical composition and metallurgical structure. Even if it is possible to use the same batch of material as that from which the plant is made, it is likely that the surface presented to the environment will be different from that exposed in the plant (e.g. the probe will be a machined cylinder, whereas the exposed surface of the plant will be the as-received surface of the plate).
- The stress state of the probe will be different from that of the plant.
- The probe will be exposed to only one sample of the range of environments to which the plant is exposed. This is particularly important in multi-phase systems, where completely different behaviours can be expected for surfaces exposed in the different phases or mixtures of phases. It is not always clear that two phases are present. Thus hydrocarbon pipelines may contain traces of water that will drop out in low regions of the pipe, or possibly separate out onto hydrophilic surfaces (such as the oxidised region adjacent to welds). The regions in contact with water will typically be the sites of corrosion problems, and it may be very difficult to install probes that will reliably detect these problems.
- The probe will typically experience different flow conditions from the plant. Very often the probe will create a flow disturbance; this may lead to an increased corrosion rate, but equally it may inhibit pitting corrosion.
- The area of the probe is usually much smaller than that of the plant, and many monitoring methods require that the probe is electrically isolated (and there are questions about galvanic effects if the probe is not isolated). This has important implications for localised corrosion. First, initiation of pits and other localised corrosion phenomena is typically a relatively rare event, and it will therefore be less likely in any given period on a small probe than on a large area of the plant, meaning that the plant will suffer the localised corrosion before it is revealed by the probe (assuming of course that a monitoring method capable of detecting the initiation of localised corrosion is used). Additionally, stable pitting or crevice corrosion depends on a relatively large cathode area to supply the current to the active pit or crevice. The small probe may not have a sufficient area to do this, so that stable pitting cannot be achieved, whereas it can for pits that are driven by the much larger area of the plant.

Based on these limitations, we must think carefully about the value of probes for corrosion monitoring. Probes should be regarded as a method of determining the corrosivity of the environment, rather than measuring the actual corrosion of the plant, thereby providing information for the management of inhibitor additions or a warning of upset conditions that result in an increased corrosivity. Consequently, there is an argument that corrosion probes should be designed to be slightly less corrosion-resistant than the plant material, so that there is a reasonable expectation that they will provide an early warning of problems, rather than responding only after corrosion problems have started on the plant. However, this leads to a delicate balancing act; as an example, we might consider using Type 304 stainless steel as a probe for the monitoring of a plant constructed out of Type 316. This then implies that we are going to operate the plant in conditions where 304 will not normally suffer corrosion. This should certainly avoid corrosion problems with the plant, but it will not make full use of the capabilities of the 316. We therefore need a probe material that is closer to 316 in composition, something like a 'lean' 316, with chromium and/or molybdenum concentrations just below the specification minimum. However, this then leads to further questions, as the material would have to be a specialist cast; it would probably be processed rather differently from a commercial cast, and it would probably have a quite different concentration and size distribution of MnS inclusions. As the latter are well known to be very important in the initiation of pitting corrosion, this leads to uncertainty as to whether the probe is more or less sensitive to corrosion than the plant, and by how much.

It seems unlikely that corrosion monitoring will be able to do without the use of probes for the foreseeable future, but careful thought needs to be given to the implications of the factors indicated above for the reliability of the data obtained.

1.4   Accuracy of electrochemical methods

Electrochemical methods, such as the linear polarisation resistance method, are popular for corrosion monitoring, offering the advantage of an instantaneous, online measure of corrosion rate. These methods mostly depend on the Stern-Geary equation, the derivation of which makes a number of assumptions:

- There is one anodic metal dissolution reaction and one cathodic reduction reaction (oxygen reduction or hydrogen evolution)
- Both anodic and cathodic reactions are far from equilibrium, so that the reverse reaction can be ignored
- Both reactions obey Tafel's Law ($\log i \propto E$, hence $i \propto \mathrm{d}i/\mathrm{d}E$)
- The solution resistance is negligible (this can be measured and corrected for using electrochemical impedance spectroscopy if it presents a problem)
- The measurement is made sufficiently slowly that capacitive currents can be neglected.

For many (probably most) real systems, these assumptions are not valid; metal dissolution is usually not under pure activation control and there will be surface films of oxide, inhibitor or other absorbed species. Indeed, if the anodic reaction is under pure activation control, then it is almost inevitable that the alloy will be corroding too fast to be practically useful (except in a few particularly borderline systems such as steel in de-aerated water at ambient temperature). Similarly, the cathodic reaction is often under diffusion or mixed control (e.g. oxygen reduction); while complete diffusion control (when the oxygen reduction reaction is at the limiting current density) can be approximated by an infinite Tafel slope, mixed control (which is probably more common in reality) does not provide the required behaviour, and will lead to errors. Additionally, there may be other, competing reactions; thus copper will usually be corroding in conditions where the reverse copper deposition reaction occurs at a significant rate. A related effect is found with nickel-base alloys in hydrogenated high-temperature water (i.e. pressurised water reactor environments) where the kinetics of the hydrogen/water redox reaction are faster than the kinetics of the corrosion reaction, so that electrochemical measurements do not provide any information about the corrosion rate.

The harmonic analysis and electrochemical frequency modulation methods are interesting, in that, according to the theory on which they are based, they permit the independent determination of the B-value for the Stern-Geary equation. However, it must be appreciated that the theory still depends on the same assumptions; both methods will produce a B-value for any system, but this does not prove that the B-value is valid. If one or other of the anodic and cathodic reactions does not obey Tafel's Law, the Stern-Geary equation is not applicable, so the calculated B-value is meaningless. It may be that the resultant estimate of corrosion rate is not too far from the true value ('too far' in corrosion monitoring is an interesting concept – within a factor of two or three appears to be doing quite well, and may well be acceptable, especially if the response to change in rate is detected reliably).

1.5   Epilogue and answers to the questions

The answer to all of the questions in the Introduction is, of course, no. Simply by monitoring corrosion, we do not change the corrosion rate, the cost of corrosion, the hazards of corrosion or the environmental impact of corrosion. This does not mean that corrosion monitoring is of no value, but it does mean that it is only of value when it forms an *input* to an active *corrosion management* programme that has *outputs* that will lead to some or all of the above benefits. Corrosion monitoring that does nothing more than fill filing cabinets or database storage space might as well not have been measured (except that it may help to explain why that 'unexpected' corrosion failure occurred).

2

Introduction to different electrochemical corrosion monitoring techniques

Rik-Wouter Bosch and Walter F. Bogaerts

SCK·CEN, Boeretang 200, B-2400 Mol, Belgium
rbosch@sckcen.be

2.1 Introduction

The corrosion of a metal in an aqueous solution is an electrochemical process involving anodic oxidation of the metal and cathodic reduction of species from the solution. We distinguish two types of corrosion: uniform corrosion and localised corrosion. For uniform corrosion, the corrosion rate is usually expressed as the reduction in thickness of a particular metallic sample in millimetres per year. For localised corrosion, the corrosion rate, expressed as the total amount of material loss, is not very meaningful and therefore not very important. The aim is to detect the localised corrosion event itself such as pitting corrosion or stress corrosion cracking (SCC).

To monitor such corrosion phenomena, several techniques are available. Among others these techniques can be divided into the following two categories:

- Electrochemical techniques
- Non-electrochemical techniques.

An overview of available corrosion monitoring techniques is shown in Table 2.1. This selection is based on some overview papers and based on the authors' experiences [1–4]. We will not discuss all monitoring techniques, but will focus on electrochemical techniques. For each monitoring technique we will give a short description, the principle, and some applications.

2.1.1 Electrochemical techniques

The most commonly used electrochemical technique is the Linear Polarisation Resistance (LPR) technique. Other electrochemical techniques that are available to determine a corrosion rate are Electrochemical Impedance Spectroscopy (EIS) and Tafel extrapolation [5–7]. For instantaneous corrosion rate measurements, the LPR and EIS techniques can be used if the anodic and cathodic Tafel parameters are known. The Tafel extrapolation technique permits determination of the corrosion rate and the Tafel parameters, but in most cases, it is not suitable for instantaneous corrosion rate measurements because the system is polarised over a wide potential range such that the measurement is time consuming and the electrode surface is affected during the measurement. Special categories are the non-linear distortion techniques such as Harmonic Analysis (HA) [8–11] and the electrochemical frequency modulation (EFM) (or intermodulation distortion) technique [12,13]. They provide a corrosion rate without prior knowledge of the Tafel parameters using only a small perturbation

5

Table 2.1 Overview of corrosion monitoring techniques

Electrochemical	Non-electrochemical
Corrosion Potential (ECP)	Electric Resistance (ER) Probe
Redox Potential	Direct Current Potential Drop (DCPD)/
	Alternating Current Potential Drop (ACPD)
Tafel Extrapolation	Field Signature Method (FSM)
Linear Polarisation Resistance (LPR)	Contact Electric Resistance (CER)
Electrochemical Impedance	Acoustic Emission (AE)
Spectroscopy (EIS)	
Harmonic Analysis (HA) and	Ultrasonic Testing (UT)
Intermodulation Distortion (EFM)	
'Electrochemical Nose' (multi wire	Hydrogen Concentration
probe sensor)	
Electrochemical Noise (potential and	
current)*	

* Will be discussed in other chapters of this book.

signal. Interpretation of the measured data is, however, not always easy. A very novel technique is the use of a so-called 'electrochemical nose' [14,15]. This technique uses an array of electrochemical sensors from which the output is interpreted by means of an artificial neural network. The outcome of the neural network is a degree of corrosiveness of the system. One of the most used techniques for monitoring localised corrosion is electrochemical noise. This is a generic term used to describe spontaneous fluctuations of potential or current, which occur at the surface of freely corroding electrodes that can be picked up by a suitable electrode arrangement [16]. This technique will not be discussed here as it receives adequate coverage in several other chapters of the book.

A major advantage of the electrochemical techniques for on-line corrosion monitoring is the rapid response time. Limitations are that a conducting electrolyte is necessary to perform the measurements, and the complications that arise with the interpretation of the measurements results.

2.1.2 Non-electrochemical techniques

For the non-electrochemical technique, the Electric Resistance (ER) measurements are most commonly used. With the ER technique, the change in resistance of a thin metallic wire or strip sensing element is measured as its cross-section decreases from metal loss due to corrosion. The corrosion rate is calculated by the loss of metal between two readings [7]. The ER technique uses the ohmic resistance as a physical parameter that has to be measured. This principle, which is simply based on Ohm's law, is used by a number of other non-electrochemical techniques as well. With the DCPD (Direct Current Potential Drop) and ACPD (Alternating Current Potential Drop) techniques, the increasing resistance of a test specimen with a growing crack is measured. The potential drop is measured with respect to a fixed current and so related to the ohmic resistance of the test specimen. The Field Signature Method (FSM) works on the same principle. According to the inventors:

The FSM detects metal loss, cracking, pitting and grooving due to corrosion by detecting small changes in the way electrical current flows through a metallic structure. With FSM, sensing pins or electrodes are distributed over the area to be monitored, with a typical distance between the pins of two to three times the wall thickness. An initial voltage measurement is taken, and subsequent changes in electrical field pattern are detected and compared against the initial measurement to infer structural changes in the monitored area. [17]

The Contact Electric Resistance (CER) technique developed by the Technical Research Centre of Finland (VTT) measures in-situ the ohmic resistance of a metal/metal oxide layer and can so be used to investigate the electronic properties of surface films formed in-situ [18]. Besides these techniques, a number of other physical based measurement methods are available. The measurement of hydrogen concentration on the outside of pipelines can be related to corrosion, assuming that the cathodic reaction is water reduction. This produces hydrogen that penetrates through the metallic structure which can be related to the corrosion rate. Ultrasonic Testing (UT) has been widely used as a non-destructive testing (NDT) technique especially to investigate welded structures. Acoustic Emission (AE) is still in the picture as a tool for SCC monitoring [19–21].

ER and LPR techniques found their way into practical applications in industrial environments. The measurement techniques have been standardised in the ASTM standard G-96-60, *On-Line Monitoring of Corrosion in Plant Equipment (Electrical and Electrochemical Methods)* [7].

2.2   Electrochemical monitoring techniques

2.2.1   Corrosion potential

The corrosion potential is a mixed potential and is a result of the balance between the anodic oxidation reactions and the cathodic reduction reactions at a metallic surface. It is a useful measure to determine whether corrosion takes place or not. For the proper interpretation of the measured values some background knowledge of the corrosion system is required. The corrosion potential is measured with a suitable reference electrode. The potential readings have to be recalculated to the Standard Hydrogen Electrode (SHE) scale to make comparison possible between data obtained with different reference electrodes. Table 2.2 shows some reference electrodes for use in ambient environments.

For the interpretation of the corrosion potential value, a Pourbaix or E–pH diagram [23] is one of the most powerful tools. In a Pourbaix diagram, the thermodynamic stability of a metal in water is given as a function of pH and potential. It allows one to determine the region where the metal is susceptible to corrosion (active

Table 2.2   Ambient temperature reference electrodes [22]

Reference electrode	Electrolyte	Potential (mV vs. SHE)
Hydrogen electrode	pH 0 and p_{H2} 1 bar	0
Silver/silver chloride electrode	0.1 M KCl	288
Silver/silver chloride electrode	3 M KCl	197
Copper/copper sulphate electrode	1.43 M $CuSO_4$ (saturated)	300

region), where there is no corrosion at all (immunity region), and where the metal is covered by a protective oxide layer (passive region). It should be mentioned, however, that even though Pourbaix diagrams are very helpful, they do not give any information on the kinetics of reactions. For instance, the corrosion rate in the passive range not only depends on the thermodynamic stability of metal oxides, but moreover on the protective nature of the passive film. Also, the effect of more complex chemistries in the surroundings, e.g. aggressive or inhibitive ions, is not considered in these diagrams.

A typical application is the monitoring of a cathodic protection system where the corrosion potential of a metallic structure is forced to be in the immunity region. According to the NACE Corrosion Reference Handbook, for steel, a potential of − 850 mV versus the saturated copper/copper sulphate reference electrode is required [24]. Alternatively, for anodic protection, a potential is chosen in the passive region of the Pourbaix diagram to maintain a protective oxide layer.

For corrosion in high-temperature water, the application is not much different to that for ambient applications. Pourbaix diagrams are frequently used to explain the meaning of the measured corrosion potentials. Pourbaix diagrams for metals in high-temperature water are however more difficult to obtain than those at room temperature. Experimental verification is also more complicated.

To measure corrosion potential, a reference electrode is needed. For measurements in high-temperature water, it is in most cases a technological challenge to develop such a reference electrode. A number of high-temperature reference electrodes developed in the last decades are summarised in Table 2.3.

Besides the use of Pourbaix diagrams to interpret the corrosion potential, other relationships with the corrosion potential are also used. An important corrosion problem in light water reactors (LWRs) is the sensitivity to SCC of Ni-based alloys and irradiated stainless steels. Both corrosion processes have a relationship with the corrosion potential. Figure 2.1 shows these relationships schematically.

It is interesting to see that for both systems, the crack growth rate depends on the corrosion potential but in a different way. The first example is (irradiated) stainless steel in a boiling water reactor (BWR). The crack growth rate decreases significantly with decreasing corrosion potential. This behaviour has resulted in the application of so-called HWC (hydrogen water chemistry). Hydrogen is added to the primary BWR circuit to lower the corrosion potential [31,32]. The second example is the behaviour of Ni-based alloys in high-temperature water. There is a particular region where the

Table 2.3   High-temperature reference electrodes and their potentials at 300°C [25–30]

Reference electrode	Electrolyte	Potential (mV vs SHE)
Hydrogen electrode	pH 0 and p_{H2} 1 bar	0
Internal silver/silver chloride electrode	0.1 M KCl	7
EPBRE (External Pressure Balanced Reference Electrode) silver/silver chloride electrode	0.1 M KCl	−37
Yttrium stabilised zirconium electrode	Ni/NiO	−783 (pH 7.3)
Yttrium stabilised zirconium electrode	Cu/CuO	−1010 (pH 7.3)
Yttrium stabilised zirconium electrode	Fe/Fe$_3$O$_4$	−473 (pH 7.3)

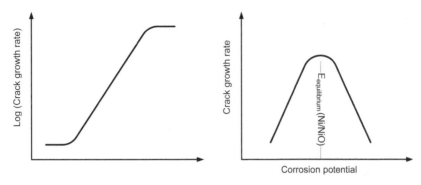

2.1 Crack growth as a function of corrosion potential. Left-hand side: a susceptible stainless steel in high-temperature water; right-hand side: a susceptible Ni-alloy in high-temperature water

crack growth rate is at its maximum. This potential range should be avoided but is by accident close to the corrosion potential (ECP) of the primary pressurised water reactor (PWR) water chemistry. Notice that the crack growth rate of the first example covers several orders of magnitude. This is not the case for the second example where the variation of crack growth is within one order of magnitude. In Ref. 32, Andresen combines both curves to obtain a general K versus crack growth behaviour.

The challenges in ECP measurements are related to the development of reference electrodes that can survive extreme conditions such as in-core LWR conditions, super critical water systems and various chemical process conditions with aggressive solutions at high temperature. Figure 2.2 shows a prototype miniature Yttrium Stabilised Zirconia (YSZ) reference electrode for use in high-temperature water [33,34].

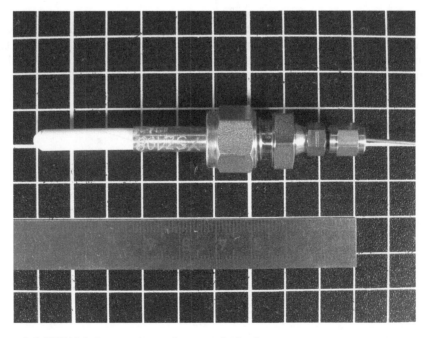

2.2 YSZ high-temperature reference electrode

Alternatively, corrosion sensors that have a (very) long lifetime, for instance, to monitor buried pipelines, tanks or concrete structures, are important as well. The second step in all corrosion potential monitoring systems is the proper interpretation of the results.

2.2.2   Tafel extrapolation

Polarisation is described as the extent of the change in potential of an electrode from its equilibrium potential caused by a net current flow to or from the electrode. The relationship between the current and potential of an electrode can be described by a polarisation curve. Consider a corrosion process where both the anodic and cathodic processes are activation controlled, the current–potential relationship is then given by

$$i = i_{corr}\left[\exp\left((\ln 10)\frac{\eta}{b_a}\right) - \exp\left(-(\ln 10)\frac{\eta}{b_c}\right)\right] \qquad [2.1]$$

where i_{corr} is the corrosion current density, η the overpotential, b_a the anodic Tafel parameter and b_c the cathodic Tafel parameter. This equation follows from the mixed potential theory, first described by Wagner and Traud [35]. It is also common to use the inverse function of equation 2.1, resulting in the Tafel equations [36]. The anodic Tafel equation for overpotentials larger than the corrosion potential

$$\eta = b_a \log\left(\frac{i_a}{i_{corr}}\right) \qquad \text{for } \eta >> E_{corr} \qquad [2.2]$$

and the cathodic Tafel equation for overpotentials smaller than the corrosion potential

$$\eta = -b_c \log\left(\frac{i_c}{i_{corr}}\right) \qquad \text{for } \eta << E_{corr} \qquad [2.3]$$

where i_a is the anodic current density and i_c the cathodic current density of the corrosion reaction. A Tafel plot is performed on a metal specimen by polarising the specimen about 250 mV anodically and cathodically from the corrosion potential. The resulting current is plotted on a logarithmic scale as shown in Fig. 2.3. The corrosion

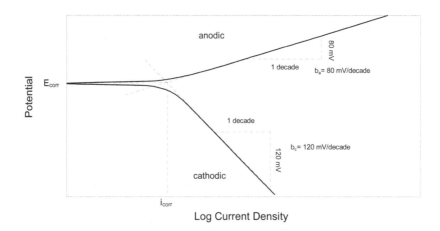

2.3 Tafel plot with extrapolation to determine the corrosion rate

current is obtained from a Tafel plot by extrapolating the linear portion of the curve to E_{corr}, also illustrated in Fig. 2.3. The slope of the anodic part of the Tafel plot is the anodic Tafel parameter (often expressed in mV/decade) and the slope of the cathodic part of the Tafel plot is the cathodic Tafel parameter.

For corroding systems that do not follow Tafel behaviour, two limiting cases can still be described: complete diffusion control of the cathodic reaction, $b_c \to \infty$, and complete passivation of the anodic reaction, $b_a \to \infty$. For corroding systems that do not follow Tafel behaviour more complex modelling of the polarisation curve is needed. A common situation is a corroding system with the cathodic reaction under mixed control, i.e. both the charge transfer and the diffusion determine the reaction rate of the cathodic reaction [37–39]. The equation for this corroding system can be described by [40]

$$i = i_{corr}\left[\exp\left((\ln 10)\frac{\eta}{b_a}\right) - \frac{\exp\left(-(\ln 10)\frac{\eta}{b_c}\right)}{1 - \frac{i_{corr}}{i_l} + \frac{i_{corr}}{i_l}\exp\left(-(\ln 10)\frac{\eta}{b_c}\right)}\right] \qquad [2.4]$$

where i_l is the limiting current density of the cathodic reaction. The measurement of a polarisation curve is standardised by the ASTM standard G 5-94, *Standard Reference Test Method for Making Potentiostatic and Potentiodynamic Anodic Polarisation Measurements* [41]. With Tafel extrapolation the corrosion rate and both the anodic and cathodic Tafel parameters can be obtained, but in most cases, it is not suitable for instantaneous corrosion rate measurements because the system must be polarised over a wide potential range such that the measurement is time consuming and the electrode surface is affected during the measurement.

2.2.3 Linear polarisation resistance

With the LPR technique, a polarisation resistance is obtained from a corroding system [42–46]. In the vicinity of the corrosion potential, the corroding system is polarised over a small potential range (-10 to $+10$ mV) and the current is measured. This results in a linear relationship between the potential and the current. When graphically plotted, the slope of this graph $\frac{dE}{di}$ is a resistance, called polarisation resistance. Figure 2.4 shows the principle.

With the polarisation resistance, the corrosion rate can be calculated with the Stern–Geary equation [47]

$$i_{corr} = \frac{b_a b_c}{\ln 10 (b_a + b_c) R_p} \qquad [2.5]$$

where i_{corr} is the corrosion current density and R_p the polarisation resistance. To obtain a corrosion rate with LPR, the Tafel parameters are needed. They can be determined with Tafel extrapolation or taken (if available) from the literature. Complications for the LPR arise when the electrolyte conductivity is low [48]. The extra ohmic resistance that exists between the tip of the reference electrode and the working electrode leads to an underestimation of the corrosion rate. Compensation for the ohmic resistance is possible by feed backward compensation, if the ohmic resistance is known, or by a current interrupt technique, if a proper reference electrode is used.

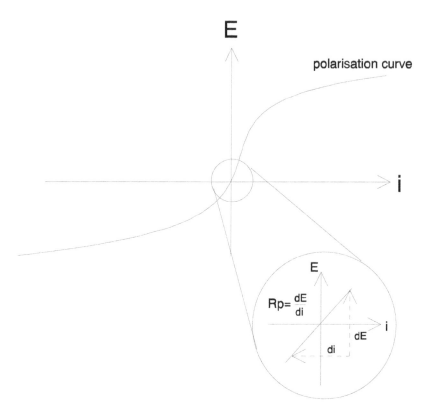

2.4 Determination of the polarisation resistance

Corroding systems that do not follow Tafel behaviour require a more complex modelling of the polarisation resistance than is achieved with the Stern–Geary equation. For example, under activation control of the anodic reaction and mixed control of the cathodic reaction, the expression for R_p can be given by [49]

$$\frac{1}{R_p} = 2.3i_k \left(\frac{1}{b_a} + \frac{1}{b_c} \left[\frac{1}{1 + \dfrac{i_o}{i_{d,c}} \exp\left(-\dfrac{\alpha_c z_c F}{RT}\left(E_k - E_{o,c}\right)\right)} \right] \right) \qquad [2.6]$$

where i_k is the corrosion current density, i_o the exchange current density and $i_{d,c}$ the limiting current density for the cathodic reaction, α_c the transfer coefficient and Z_c the number of electrons involved in the cathodic reaction, E_k the corrosion potential and $E_{o,c}$ the reversible cathodic potential. LPR measurements are also standardised and described by ASTM standard G 59-91, *Standard Practice for Conducting Potentiodynamic Polarisation Resistance Measurements* [50].

For many practical applications, the anodic and cathodic Tafel parameters are put together as a B value:

$$B = \frac{b_a b_c}{\ln 10\left(b_a + b_c\right)} \qquad [2.7]$$

$$i_{corr} = \frac{B}{R_p}$$ [2.8]

These B values are available for a number of systems. Table 2.4 summarises a few examples taken from the *NACE Corrosion Engineer's Reference Book* [51].

2.2.4 Electrochemical impedance spectroscopy

With EIS, a small sinusoidal (potential) perturbation of a certain frequency is used to distort an electrochemical system. The (current) response is measured and the modulus ($|E|/|i|$) and phase of the impedance are determined [52–55]. With EIS, information regarding electrochemical corrosion phenomena can be obtained in-situ, without the use of a reference electrode and with only a small perturbation signal. The latter means that this technique is non-destructive and a good candidate for on-line measurements.

Proper experimental data can be obtained from a corroding system if three requirements are met: (1) linearity, (2) causality and (3) stability [53]. The first requirement is met by using a small potential perturbation so that the system stays in the linear region. The second requirement is generally not a problem as long as the background noise is small. The third requirement is more difficult to satisfy as a corroding system is always changing with time. If these changes are slow, the system is assumed to be quasi-stable, and good experimental data can be obtained.

Figure 2.5 shows the principle of this technique. Here the electrochemical system is presented as a function between the current i and the potential E. The amplitude of the potential perturbation must be small in order not to exceed the linear region of the current–potential relationship in the neighbourhood of the open circuit (corrosion) potential or the applied dc-potential. Figure 2.6 shows the principle of a sinusoidal potential perturbation at an arbitrary dc potential. If the amplitude is too large, the linear region is abandoned. The response will contain non-linear elements such as a rectification current and higher harmonics, as will be discussed in section 2.2.5.

The measurements are performed over a wide frequency range and the results can be presented in a Bode diagram (modulus $|Z|$ and phase shift φ versus frequency) or in a Nyquist diagram (the real part of the impedance versus the imaginary part of the impedance). From the potential perturbation ΔE, current response Δi, and phase shift φ, the real and imaginary parts of the impedance can be calculated.

$$Re\,alZ = \frac{\Delta E}{\Delta i}\cos\varphi \qquad\qquad Im\,agZ = \frac{\Delta E}{\Delta i}\sin\varphi$$ [2.9]

Table 2.4 B-values for several corrosion systems

Corroding system	B (mV)
Iron in 1 N H$_2$SO$_4$	10–20
Carbon steel in seawater	25
Stainless steel 304L in 1 N H$_2$SO$_4$, oxygenated	22
Alloy 600 in lithiated water, 288°C	24
Zircaloy 2 in lithiated water, 288°C	81

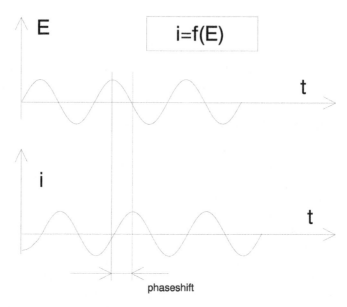

2.5 Principle of electrochemical impedance spectroscopy, sinusoidal potential perturbation with current response and phase shift

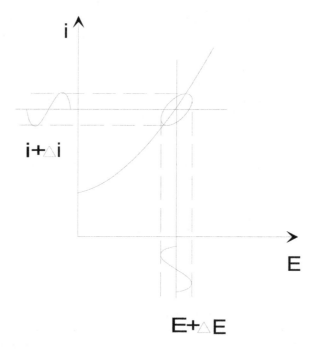

2.6 Polarisation curve with perturbation signal ΔE and response Δi at an arbitrary dc potential E and corresponding dc-current i

An example of a Nyquist diagram is shown in Fig. 2.7. The polarisation resistance and ohmic resistance can be determined graphically. The ohmic resistance is the high frequency intersection with the real axis, the polarisation resistance is the low frequency intersection with the real axis. The frequency in Fig. 2.7 is marked as ω, the angular frequency in rad/s.

The polarisation resistance is then obtained according to the following definition [56,57]

$$R_p = |Z(j\omega)|_{\omega \to 0} - |Z(j\omega)|_{\omega \to \infty} \qquad [2.10]$$

where R_p is the polarisation resistance, $|Z(j\omega)|_{\omega \to 0}$ is the low frequency limit of the impedance and $|Z(j\omega)|_{\omega \to \infty}$ is the high frequency limit of the impedance. To obtain the high frequency limit of the impedance is generally not very difficult. To determine the low frequency limit, however, can give problems as it is not always clear when the low frequency limit has been reached. If the polarisation resistance is obtained, the corrosion rate can be calculated with the Stern–Geary equation 2.5. As with EIS, the ohmic resistance is determined; EIS is particularly useful for corrosion rate measurements in low-conductivity media.

Interpretation of the impedance results can be done by means of an equivalent electrical circuit representing the metal–solution interface [58]. An equivalent circuit is constructed by resistors and capacitors that represent the different electrochemical processes that can occur at the electrode. An example of a simple corrosion system where both the anodic and cathodic reactions are controlled by charge transfer is shown in Fig. 2.8.

The experimental data of the EIS measurements are fitted with an equivalent circuit to obtain the different circuit elements. Although it is possible to obtain a perfect fit between most of the experimental EIS data and an equivalent circuit, it is sometimes difficult to relate the circuit elements with the corresponding physical and chemical processes at the corroding interface.

A more rigorous approach is the modelling of the EIS data with a transfer function [54,55,59–61]. Such a function is obtained by solving the appropriate differential equation describing the electrochemical (corroding) interface. For example, if the

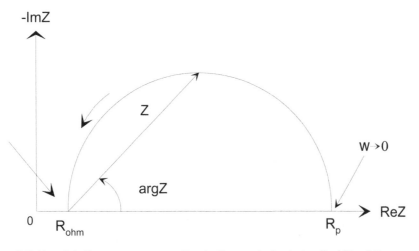

2.7 Nyquist diagram corresponding to the equivalent circuit of Fig. 2.8

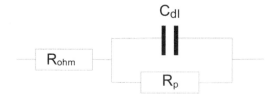

2.8 Randles equivalent circuit representing a simple corroding system with ohmic resistance R_{ohm}, polarisation resistance R_p and double layer capacity C_{dl}

current response is dependent on the potential and concentration fluctuations (ΔC), the current response follows from

$$\Delta i = \frac{di}{dE}\Delta E + \frac{di}{dC}\Delta C \qquad [2.11]$$

where C is the concentration and ΔC the concentration fluctuation of the species involved in the electrode reactions. This equation is solved in the Laplace (frequency) domain and used to fit the experimental data, to obtain the parameters of interest.

Computerised acquisition of impedance data can be carried out with a variety of equipment. For the low frequency measurements, data can be obtained by Fourier Transformations. Electronic equipment such as a Frequency Response Analyser (FRA) or Lock-In Amplifier is available for measurements up to 1 MHz.

Two examples will be given here. The first example is the use of EIS to study localised corrosion.

EIS to monitor SCC [62]

Impedance measurements have been carried out during a slow strain rate test (SSRT). The phase difference between a non-stressed and stressed sample is measured. The difference in phase is related to the surface morphology, i.e. smooth, cracked or branched surface. Figure 2.9 shows the evolution of the phase shift during the SSRT test. The phase shift starts to increase after a certain time suggesting the formation of SCC cracks. Figure 2.10 shows SEM pictures of the fracture surface at two magnifications. It can be clearly seen that the specimen was broken due to transgranular SCC.

High-temperature applications [63,64]

A typical EIS application is the monitoring of oxide film properties in high-temperature water. The following example results from tests with stainless steel in PWR primary water (300°C, 150 bar, 2 ppm Li, 400 ppm B and 25 cc/kg H_2). Figure 2.12 shows a SEM picture of the surface after exposure during a 4-week period. A porous spinel type layer is formed. For the interpretation of the impedance data, the equivalent circuit shown in Fig. 2.11 is proposed.

Figure 2.12 shows a Nyquist diagram after 2 weeks of testing with a fit result based on the equivalent circuit of Fig. 2.11. A reasonably good fit is obtained. The fit result can be improved significantly when the number of elements in the equivalent circuit model is increased. However, care should be taken not to add circuit elements without a sound physical basis. Therefore, we stick to this model realising that

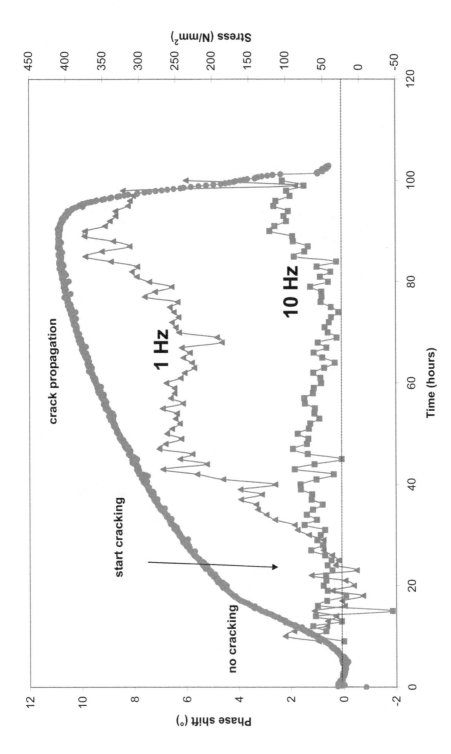

2.9 Crack initiation and propagation are related to the evolution in phase shift during a SSRT test of solution annealed Type 304 SS in 0.01 M Na$_2$SO$_4$ at 300°C

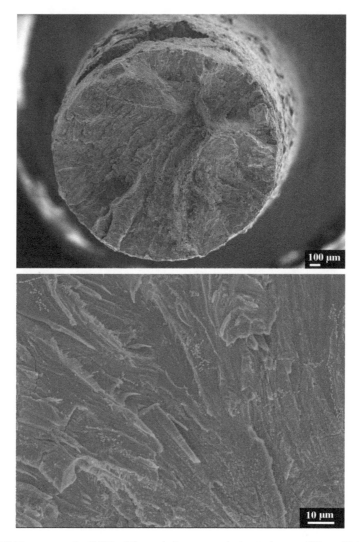

2.10 Transgranular SCC of the solution annealed specimen of Type 304 SS in 0.01 M Na$_2$SO$_4$ at 300°C

improvements are still possible, but that additional information is required about the nature of the oxide film and the electrochemical processes that take place. Figure 2.13 shows the evolution of the CPE exponent n with the in- and out-of-core data. It shows that there is a clear difference between the morphology of the oxide layer in the in- and out-of-core conditions. The low value of n for the in-flux data suggests a more porous structure of the oxide film.

2.2.5   Harmonic analysis

Due to the non-linear nature of the current–potential relationship of an electrochemical system, harmonics are generated when a sinusoidal potential perturbation is applied. As shown in Fig. 2.14, a small potential perturbation results in a linear

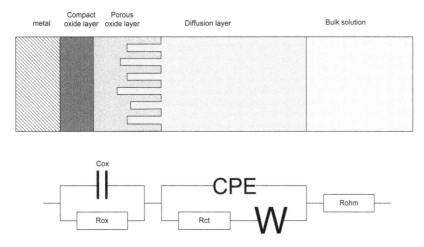

2.11 Equivalent circuit of the metal/oxide/solution interface of stainless steel in high-temperature water; C_{ox} is the capacity of the oxide layer; R_{ox} is the resistance of the oxide layer; CPE is the Constant Phase Element associated with the electrochemical double layer, R_{ct} the charge transfer resistance, W the Warburg impedance and R_{ohm} the electrolyte resistance

current response, a large potential perturbation results in a non-linear current response.

Analysis of the harmonics of the ac current response gives the possibility to obtain kinetic information of the electrode processes.

Emphasising an electrochemical corrosion process, analysis of the harmonics of the ac current response gives the possibility of obtaining the corrosion rate and both the anodic and cathodic Tafel parameters. Theoretical work by Dévay, Mészáros and colleagues resulted in mathematical expressions for the corrosion current and Tafel parameters [65,66]. For small values of the amplitude of the potential signal and for $\eta = 0$, this resulted in the following equations for the corrosion rate (i_{corr}) and Tafel parameters (β_a) and (β_c) (for $\beta_a < \beta_c$)

$$i_{corr} = \frac{i_1^2}{\sqrt{48\left(2i_1i_3 - i_2^2\right)}} \qquad [2.12]$$

$$\frac{1}{\beta_a} = \frac{1}{2U_o}\left(\frac{i_1}{i_{corr}} + 4\frac{i_2}{i_1}\right) \qquad [2.13]$$

$$\frac{1}{\beta_c} = \frac{1}{2U_o}\left(\frac{i_1}{i_{corr}} - 4\frac{i_2}{i_1}\right) \qquad [2.14]$$

where i_1, i_2 and i_3 are the first, second and third harmonic current components taken from the frequency spectrum of the ac current response. The harmonic equations 2.12 through 2.14 have been used by different authors to measure corrosion rates in acid media with and without the addition of inhibitors [67–72]. Lawson and Scantlebury used harmonic analysis (HA) to monitor steel rebar corrosion in concrete [73]. Mészáros et al. measured the rate of underpaint corrosion with HA [74]. For a

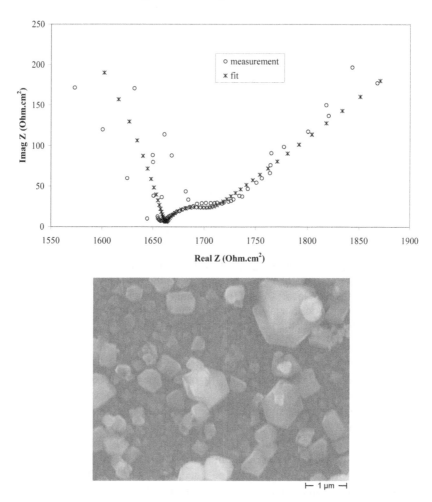

2.12 Measured and simulated Nyquist diagram after 2 weeks of testing of a stainless steel 304 sample + a SEM picture of the structure of the oxide layer

corroding system under mixed control, i.e. both the charge transfer and diffusion contribute to the total cell polarisation, equations have been derived for the first, second and third harmonic current. Experimental results on mild steel in an aerated sodium sulphate solution show that it is possible to obtain corrosion rates that are in agreement with the Tafel extrapolation technique [75]. An application in the nuclear field is the measurement of corrosion rates of pure copper in a bentonite/groundwater environment by Rosborg et al. [76].

A different approach to determine the corrosion rate from the non-linear response due to a sinusoidal potential perturbation is Harmonic Impedance Spectroscopy (HIS) [77–79]. The higher harmonic currents are transformed to impedances, having the advantage that the first harmonic impedance can be analysed by an equivalent circuit. This allows stripping off all the non-relevant components (such as the ohmic resistance and the double layer capacity) of the first harmonic impedance. Finally, a set of three equations similar to equations 2.12 through 2.14 have to be solved

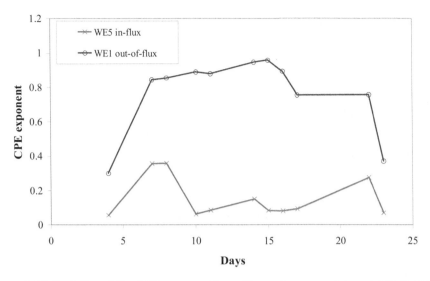

2.13 Evolution of the CPE exponent from fit over frequency interval 100–0.01 Hz

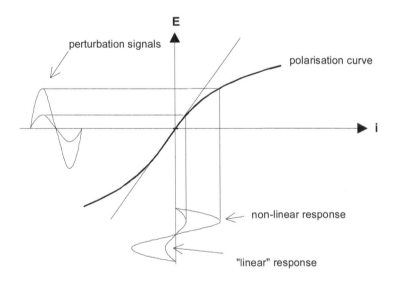

2.14 A polarisation curve and non-linear response due to a sufficiently large sinusoidal potential perturbation

to obtain the corrosion rate and Tafel parameters. HIS has been used to measure corrosion rates in cathodically protected systems.

HA has also been used in a more phenomenological approach. Kendig and Anderson have used HA to investigate the initiation of pitting corrosion [80]. A sudden increase in the non-linear behaviour of a corroding system, reflected by the higher harmonics in the frequency spectrum, indicates the initiation of pitting corrosion. Kihira et al. developed an instrument to detect the quality of the protective properties of rust films on weathering steel [81]. In addition to the ac-impedance, harmonic currents could also be measured. The latter shows that there is a non-linear current–potential relationship, indicating a corrosion resistance based on charge transfer.

2.2.6   Intermodulation distortion technique

With the Intermodulation Distortion technique, also called EFM (Electrochemical Frequency Modulation), a potential signal consisting of two sine waves of different frequencies is applied to a non-linear system. The ac-response contains non-linear components at harmonic and intermodulation frequencies. Analysis of these components can give information about the behaviour of the system under investigation. Mészáros and Dévay have made a theoretical study, like their work on HA, for the EFM technique [82]. Bosch and Bogaerts have further elaborated on this and explored the possibilities for uniform corrosion rate measurements [83,84].

Principle of the EFM technique

With the EFM technique, a potential perturbation by two sine waves of different frequencies is applied to a corroding system. The ac-current response due to this perturbation consists of current components of different frequencies. As a corrosion process is non-linear in nature, responses are generated at more frequencies than the frequencies of the applied signal. Current responses can be measured at zero, harmonic and intermodulation frequencies. The principle of the EFM technique is illustrated in Fig. 2.15.

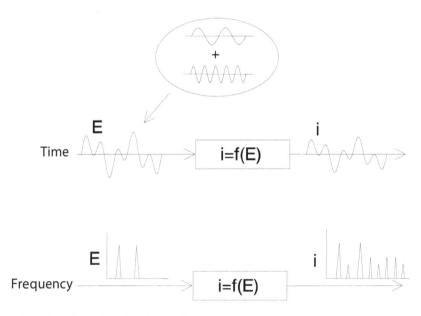

2.15 Principle of the EFM technique

They defined a potential signal U of two frequencies ω_1 and ω_2 and two different amplitudes U_1 and U_2

$$U = U_1\sin(\omega_1 t) + U_2\sin(\omega_2 t) \qquad [2.15]$$

For a corroding system that follows Tafel behaviour at the corrosion potential, for small values of U_1 and U_2, the following equations were derived for the corrosion rate and Tafel parameters

$$i_{corr} = \frac{U_2}{U_1}\frac{i_{\omega_1}^{\,2}}{2\sqrt{8i_{\omega_1}i_{\omega_1\pm2\omega_2} - 3\left(i_{\omega_1\pm\omega_2}\right)^2}} = \frac{U_1}{U_2}\frac{i_{\omega_2}^{\,2}}{2\sqrt{8i_{\omega_2}i_{\omega_2\pm2\omega_1} - 3\left(i_{\omega_1\pm\omega_2}\right)^2}} \qquad [2.16]$$

$$\frac{1}{\beta_a} = \frac{1}{2U_1}\left(\frac{i_{\omega_1}}{i_{corr}} \pm 2\frac{U_1}{U_2}\frac{i_{\omega_1\pm\omega_2}}{i_{\omega_1}}\right) = \frac{1}{2U_2}\left(\frac{i_{\omega_2}}{i_{corr}} \pm 2\frac{U_2}{U_1}\frac{i_{\omega_1\pm\omega_2}}{i_{\omega_2}}\right) \qquad [2.17]$$

$$\frac{1}{\beta_c} = \frac{1}{2U_1}\left(\frac{i_{\omega_1}}{i_{corr}} \mp 2\frac{U_1}{U_2}\frac{i_{\omega_1\pm\omega_2}}{i_{\omega_1}}\right) = \frac{1}{2U_2}\left(\frac{i_{\omega_2}}{i_{corr}} \mp 2\frac{U_2}{U_1}\frac{i_{\omega_1\pm\omega_2}}{i_{\omega_2}}\right) \qquad [2.18]$$

The upper signs of equations 2.17 and 2.18 refer to the case where $\beta_a < \beta_c$, the lower signs refer to the case where $\beta_a > \beta_c$, i_{ω_1} and i_{ω_2} are harmonic currents at frequency ω_1 and ω_2, $i_{\omega_1\pm\omega_2}$ is the intermodulation current at $\omega_1\pm\omega_2$ and $i_{\omega_2\pm2\omega_1}$ is the intermodulation current at $\omega_2\pm2\omega_1$.

Practical application of the EFM technique was first carried out by Bosch and Bogaerts [82,83]. Corrosion rates were measured for mild steel in an acidic environment with and without inhibitors. The corrosion rates measured with the EFM technique were in good agreement with more established techniques for corrosion rate measurement such as LPR, Tafel and EIS. For corrosion rate measurements in neutral environments, the theory for the EFM technique was extended, and experiments were carried out in different neutral solutions. Reasonably accurate corrosion rates were obtained, although the modelling of the intermodulation response can get quite complicated. This might hamper routine application as shown in a critical review of the technique by Kuş and Mansfeld [84].

A schematic of a frequency spectrum of an EFM current response, showing the different harmonic and intermodulation frequencies, is shown in Fig. 2.16.

A result obtained from a real corrosion system, i.e. mild steel in sulphuric acid solution, is shown in Fig. 2.17, using a perturbation signal with an amplitude of 20 mV for both frequencies and perturbation frequencies of 0.2 and 0.5 Hz. The choice for the frequencies of 0.2 and 0.5 Hz was based on three arguments. First, the harmonics and intermodulation frequencies should not influence each other. Second, the frequency should be as low as possible to avoid the influence of the capacitive behaviour of the electrochemical double layer. Third, the frequency should be as large as possible to reduce the time needed to perform a measurement. While these arguments do not allow the selection of 'perfect' frequencies, the chosen frequencies were considered a reasonable compromise.

The harmonic and intermodulation peaks are clearly seen and are much larger than the background noise. Analysis of the peaks at the intermodulation frequencies results in the corrosion rate and Tafel parameters. For this example, a corrosion current of 5E-4 A/cm² (5.8 mm/year) has been obtained. Figure 2.18 shows an example for the monitoring of corrosion inhibition [85]. There is good agreement among the different techniques used.

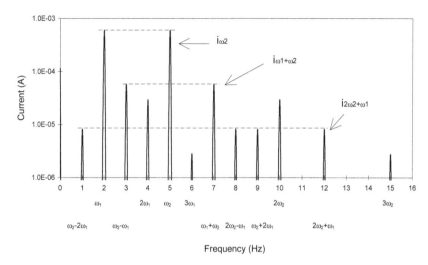

2.16 Harmonic and intermodulation frequencies resulting from the basic frequencies 2 and 5 Hz

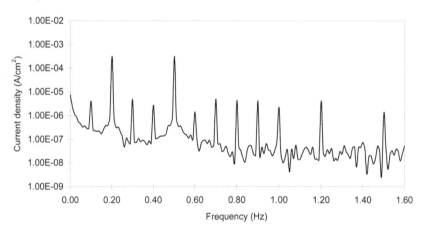

2.17 Frequency spectrum of the current response of mild steel in 0.05 M H_2SO_4

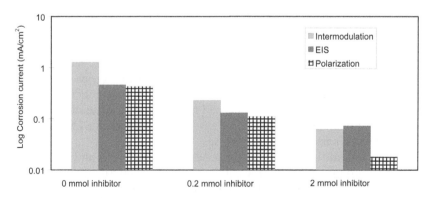

2.18 Comparison of corrosion rates obtained with the EFM technique, EIS, and polarisation curves (Tafel extrapolation) for mild steel in 0.5 M H_2SO_4 with different concentrations of inhibitor

Kendig and Anderson have used HA to investigate the initiation of pitting corrosion [80]. With EFM, a similar approach can be followed. Some attempts have been made to establish whether this is possible or not [86]. A sudden increase of the non-linear behaviour of a corroding system, reflected by the intermodulation currents in the frequency spectrum, indicates the initiation of pitting corrosion. Figure 2.19 shows a calculation example. The polarisation curve showed a large passive part and a part ($E > 0$ V) undergoing pitting corrosion. EFM spectra were calculated for two different potentials; one in the passive range ($E = -0.1$ V) and one close to the pitting potential. The spectra show a very clear difference in behaviour suggesting that this technique might be able to differentiate between a system sensitive to pitting or not.

2.2.7 'Electrochemical nose' or artificial intelligence corrosivity sensor

This section is based on work carried out by Ferreira, Bogaerts and colleagues [14,15]. 'Electronic noses' are sensor arrays with an incorporated pattern recognition system to identify patterns in odours, vapours or gases and automatically identify them. In the case of the identification of fluids, this sensor is usually called an 'electronic tongue'. Based on this concept, the idea was developed to design an electrochemical sensor for the assessment of an environment's corrosivity.

The working principle of this device relies on the potential dependence of a set of metals on environmental (i.e. fluid) characteristics such as pH, conductivity, dissolved oxygen and ion concentration (the sensing system). The values of these electrochemical potentials are the result of the influence of different combinations of these environmental parameters. In addition, the parameters that characterise an environment are a reflection of its corrosivity. Corrosion and electrochemical tests such as LPR and EIS may be performed in selected environments to give an indication of the corrosion rate. These measurements reflect the corrosivities of those particular environments. That said, if the electrochemical potential of a set of selected metals in different environments is monitored and an indication of some material's corrosion rate in these same environments is known, it should be possible to automatically identify the corrosivity of an environment using an appropriate pattern recognition system.

An example of a neural network and a possible output is shown in Fig. 2.20.

2.3 Conclusions

In addition to laboratory testing, where real service conditions are only simulated, in-situ corrosion monitoring can provide information directly from real components under real conditions. This can contribute to the economical and safe operation of nuclear plants/systems. However, it is very difficult to apply real in-situ monitoring, especially under severe conditions such as are encountered in LWRs. Therefore much research work has been carried out on different monitoring methods. Some promising (electrochemical) corrosion monitoring techniques have been briefly reviewed in this chapter.

First, some well-known techniques have been described, such as the measurement of the ECP, the LPR technique, Tafel extrapolation and EIS. Here we saw that the most interesting developments are in the application of these techniques under increasingly severe conditions (high-temperature water, up to super-critical water and in-core testing).

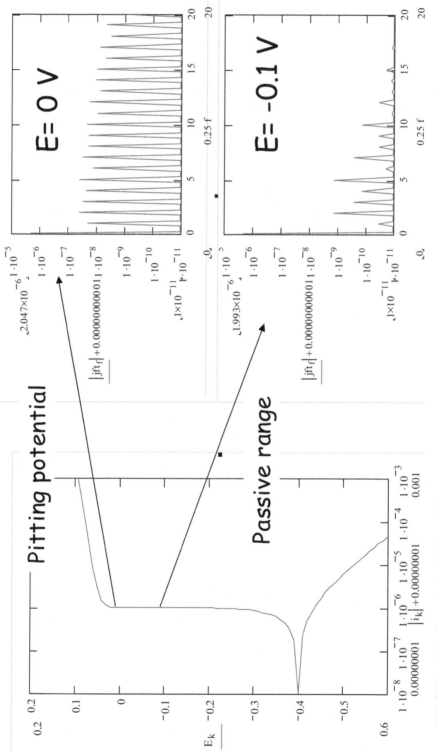

2.19 Pitting detection with the EFM technique

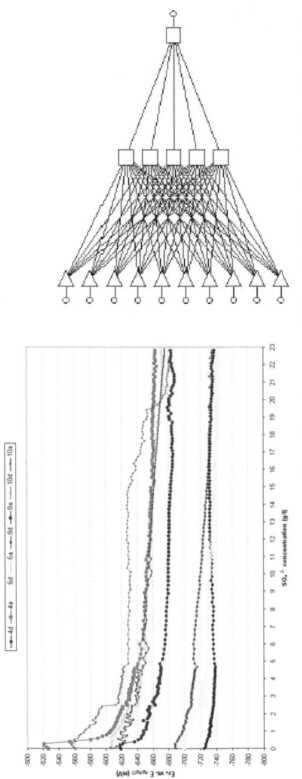

2.20 Illustration of the neural network architecture formed by 10 input neurons (7 electrode potentials, pH, conductivity and logarithm of sulphate concentration), one hidden layer with 5 hidden units and 1 output neuron (polarisation resistance values)

Some really novel techniques such as HA, EFM and 'electrochemical nose' were discussed. These techniques provide unique features in extracting information from a corroding system, but are still difficult to apply and interpret.

In reality, at this moment, technological improvements will be beneficial for existing measurement techniques, such as high-temperature reference electrodes, signal cables, signal cable feed-through, connectors and sensors that can survive severe test conditions.

References

1. J. C. Bovankovich, 'What every engineer should know about on-line corrosion monitoring', in NACE CORROSION 93, paper no. 389, 1993.
2. A. J. Perkins, 'New developments in on-line corrosion monitoring systems', in NACE CORROSION 93, paper no. 393, 1993.
3. S. W. Dean, 'Overview of corrosion monitoring in modern industrial plants', in *Corrosion Monitoring in Industrial Plants Using Nondestructive Testing and Electrochemical Methods*, ASTM STP 908, ed. G.C. Moran and P. Labine. ASTM, Philadelphia, PA, 1986, 197.
4. J. Hubrecht, R. W. Bosch, W. F. Bogaerts, J. F. Chen and P. Ferreira Gorjao, *Rec. Res. Dev. Electrochem.*, 4 (2001), 21.
5. G. L. Cooper, 'Sensing probes and instruments for electrochemical and electrical resistance corrosion monitoring', in *Corrosion Monitoring in Industrial Plants Using Nondestructive Testing and Electrochemical Methods*, ASTM STP 908, ed. G.C. Moran and P. Labine. ASTM, Philadelphia, PA, 1986, 237.
6. J. A. Beavers, N. G. Thompson and D. C. Silverman, 'Corrosion engineering applications of electrochemical techniques: laboratory testing', in NACE CORROSION 93, paper no. 348, 1993.
7. *On-Line Monitoring of Corrosion in Plant Equipment (Electrical and Electrochemical Methods)*, ASTM-Standard G-96-90, 1994 Annual Book of ASTM Standards, Vol. 03.02, ASTM, Philadelphia, PA, 1994.
8. J. S. Gill, L. M. Callow and J. D. Scantlebury, *Corrosion*, 39 (1983), 61.
9. K. Lawson and J. D. Scantlebury, *Mater. Sci. Forum*, 44&45 (1989), 387–402.
10. L. Mészáros, B. Lengyel and F. Jánászik, *Mater. Chem.*, 7 (1982), 165.
11. R. W. Bosch and W. F. Bogaerts, *J. Electrochem. Soc.*, 143 (1996), 4033.
12. R. W. Bosch and W. F. Bogaerts, *Corrosion,* 52 (1996) 205.
13. R. W. Bosch, J. Hubrecht, W. F. Bogaerts and B. C. Syrett, *Corrosion*, 57 (2001), 60.
14. P. Ferreira Gorjao, J. Hubrecht and W. F. Bogaerts, 'Environment corrosivity assessment by an artificial intelligence corrosivity sensor', in Proc. 15th International Corrosion Congress, 2002.
15. P. Ferreira Gorjao, W. F. Bogaerts, J. Hubrecht, B. Heyns and M. J. Embrechts, in EUROCORR 2003, Budapest, 2003.
16. R. Cottis and S. Turgoose, 'Electrochemical impedance and noise', in *Corrosion Testing Made Easy*, ed. B.C. Syrett, NACE International, 1999, 51.
17. http://www.roxar.com/
18. Y. Takedaa, T. Shojia, M. Bojinovb, P. Kinnunenc and T. Saario, *Appl. Surf. Sci.*, 252(24) (2006), 8580.
19. H. N. G. Wadley, C. B. Scruby and J. H. Speake, *Int. Metals Rev.*, 249(2) (1980), 41.
20. W. W. Gerberich, et al., *Mater. Sci. Eng.*, A103 (1988).
21. R. Van Nieuwenhove and R. W. Bosch, *J. Acoust. Emission*, 18 (2000), 293.
22. M. Pourbaix, 'Lectures on electrochemical corrosion', in CEBELCOR 1967, NACE International, 1995.
23. M. Pourbaix, *Atlas of Electrochemical Equilibria in Aqueous Solutions*, CEBELCOR, NACE, 1974.

24. *NACE Corrosion Engineer's Reference Book*, NACE International, 2002, 161.
25. D. D. Macdonald, T. Zhu, and X. Guan, 'Current state-of-the-art in reference electrode technology for use in high subcritical and super critical aqueous systems', in *Electrochemistry in Light Water Reactors, Reference Electrodes, Measurement, Corrosion and Tribocorrosion Issues*, ed. R.W. Bosch, D. Féron and J.-P. Celis. EFC Publications 49, Woodhead Publishing, 2007.
26. R. W. Bosch et al., 'LIRES: A European sponsored research project to develop light water reactor reference electrodes', in *Electrochemistry in Light Water Reactors, Reference Electrodes, Measurement, Corrosion and Tribocorrosion Issues*, ed. R.W. Bosch, D. Féron and J.-P. Celis, EFC Publications 49, Woodhead Publishing, 2007.
27. R. W. Bosch, W. F. Bogaerts and J. H. Zheng, *Corrosion*, 59 (2003), 162.
28. Y. J. Kim and P. L. Andresen, *Corrosion*, 59(7) (2003), 584.
29. D. D. Macdonald, S. Hettiarachchi, H. Song, K. Makela, R. Emerson and M. Ben-Haim, *J. Solut. Chem.*, 21(8) (1992), 849.
30. D. F. Taylor and C.A. Caramihas, *J. Electrochem. Soc.*, 129(11) (1982), 2458.
31. S. A. Attanasio and D. S. Morton, 'Measurement of the nickel/nickel oxide transition in Ni-Cr-Fe alloys and updated data and correlations to quantify the effect of aqueous hydrogen on primary water', in 11th Int. Conf. Environmental Degradation of Materials in Nuclear Systems (Stevenson, WA, 10–14 August 2003).
32. P. L. Andresen, *Corrosion*, 64(5) (2008), 439.
33. R.-W. Bosch, M. Wéber and W. Vandermeulen, 'Development and testing of an in-core YSZ high temperature reference electrode', in Proc. Int. Conf. on Water Chemistry in Nuclear Reactor Systems, Berlin, 2008.
34. W. Vandermeulen and R.-W. Bosch, 'Degradation of 5 mol% Y_2O_3-ZrO_2 used for the YSZ membrane electrode in high temperature water (300°C)', in Proc. Int. Conf. on Water Chemistry in Nuclear Reactor Systems, Berlin, 2008.
35. C. Wagner and W. Traud, *Z. Elektrochem. Angew. Phys. Chem.*, 7 (1938), 391–454.
36. J. Tafel, *Z. Phys. Chem.* (1905), 641–752.
37. D. Britz and P. Hougaard, *Corros. Sci.*, 23 (1983), 987–994.
38. J. G. Hines, *Br. Corros. J.*, 18 (1983), 11.
39. B. D. McLaughlin, *Corrosion*, 37 (1981), 723.
40. Z. Nagy and D. A. Thomas, *J. Electrochem. Soc.*, 133 (1986), 2013.
41. *Standard Reference Test Method for Making Potentiostatic and Potentiodynamic Anodic Polarization Measurements*, ASTM-Standard G-5-94, 1994 Annual Book of ASTM Standards, Vol. 03.02. ASTM, Philadelphia, PA, 1994.
42. L. M. Callow, J. A. Richardson and J. L. Dawson, *Br. Corros. J.*, 11 (1976), 123.
43. L.M. Callow, J. A. Richardson and J. L. Dawson, *Br. Corros. J.*, 11 (1976), 132.
44. F. Mansfeld, *Corrosion*, 29 (1973), 397.
45. F. Mansfeld and K. B. Oldham, *Corros. Sci.*, 11 (1971), 787–796.
46. F. Mansfeld, 'The polarization resistance technique for measuring corrosion currents', in *Advances in Corrosion Science and Technology*, Vol. 6, ed. M.G. Fontana and R.W. Staehle. Plenum Press, New York, 1976, 163.
47. M. Stearn and A. L. Geary, *J. Electrochem. Soc.*, 104 (1957), 56.
48. F. Mansfeld, *Corrosion*, 44 (1988), 856.
49. H. C. Albaya, O. A. Cobo and J. B. Bessone, *Corros. Sci.*, 13 (1973), 287–293.
50. *Standard Practice for Conducting Potentiodynamic Polarization Resistance Measurements*, ASTM-Standard G-59-91, 1994 Annual Book of ASTM Standards, Vol. 03.02. ASTM, Philadelphia, PA, 1994.
51. *NACE Corrosion Engineer's Reference Book*, 89.
52. D. D. Macdonald and M. C. H. McKubre, 'Electrochemical impedance techniques in corrosion science', in *Electrochemical Corrosion Testing*, ASTM STP 727, ed. F. Mansfeld and U. Bertocci. ASTM, Philadelphia, PA, 1981, 110.

53. J. R. Macdonald, *Impedance Spectroscopy, Emphasizing Solid Materials and Systems*. John. Wiley & Sons, New York, 1987.
54. C. Gabrielli, 'Use and applications of electrochemical impedance techniques', Technical Report 24, Solartron Instruments, 1994.
55. C. Gabrielli, 'Identification of electrochemical processes by Frequency Response Analysis', Technical Report 4, Solartron Instruments, 1980.
56. K. Hladky, L. M. Callow and J. L. Dawson, *Br. Corros. J.*, 15 (1980), 20.
57. S. Haruyama and T. Tsuru, *A Corrosion Monitor Based on Impedance Method, Electrochemical Corrosion Testing*, ASTM STP 727, ed. Fl. Mansfeld and U. Bertocci. ASTM, Philadelphia, PA, 1981, 167–186.
58. R. Cottis and S. Turgoose, 'Electrochemical impedance and noise', in *Corrosion Testing Made Easy*, ed. B.C. Syrett, NACE International, 1999, 35.
59. D. D. Macdonald, *Corrosion*, 46 (1990), 229.
60. U. Bertocci and R. E. Ricker, 'Impedance spectra calculated from model polarization curves', in *Electrochemical Impedance: Analysis and Interpretation*, ASTM STP 1188, ed. J.R. Scully, D.C. Silverman and M.W. Kendig. ASTM, Philadelphia, PA, 1993, 9–22.
61. U. Bertocci, 'Modeling polarization curves and impedance spectra for simple electrode systems', in *Computer Modeling in Corrosion*, ASTM STP 1154, ed. R.S. Munn. ASTM, Philadelphia, PA, 1992, 143–161.
62. R. W. Bosch, *Corros. Sci.*, 47 (2005), 125.
63. R. W. Bosch, M. Wéber and M. Vankeerberghen, *J. Nucl. Mater.*, 360 (2007), 304.
64. R. W. Bosch and M. Vankeerberghen, *Electrochim. Acta*, 52 (2007), 7538.
65. J. Dévay and L. Mészáros, *Acta Chim. Acad. Sci. Hung.*, 104 (1980), 311.
66. L. Mészáros, G. Mészáros and B. Lengyel, *J. Electrochem. Soc.*, 141 (1994), 2068.
67. J. M. Bastidas and J. D. Scantleburry, *Corros. Sci.*, 26 (1986), 341–347.
68. J. L. Dawson, J. S. Gill, I. A. Al-Zanki and R. C. Woolam, *Electrochemical Corrosion Testing Using Electrochemical Noise, Impedance and Harmonic Analysis*, Dechema-Monographs vol. 101, VCH Verlagsgesellschaft, 1986.
69. V. Lakshminarayan and S. R. Rajagopalan, 'Electrochemical Relaxation Techniques for the Measurement of Instantaneous Corrosion Rates', in Proc. NACE Corrosion, 94, 1994, 1420.
70. S. Sathiyanarayanan and K. Balakrishnan, *Br. Corros. J.*, 29 (1994), 152.
71. C. Gabrielle, M. Keddam and H. Takenouti, 'An assessment of large amplitude harmonic analysis in corrosion studies', in *Materials Science Forum*, Vol 8: *Electrochemical Methods in Corrosion Research*, ed. M. Duprat. Trans. Tech. Pub., Switzerland, 1986, 417.
72. J. S. Gill, L. M. Callow and J. D. Scantlebury, *Corrosion*, 39 (1983), 61.
73. K. Lawson and J. D. Scantlebury, *Mater. Sci. Forum*, 44&45 (1989), 387–402.
74. L. Mészáros, B. Lengyel and F. Jánászik, *Mater. Chem.*, 7 (1982), 165.
75. R. W. Bosch and W. F. Bogaerts, *J. Electrochem. Soc.*, 143 (1996), 4033.
76. B. Rosborg, J. Pan and C. Leygraf, *Corros. Sci.*, 47 (2005), 3267.
77. N. G. Thompson and B. C. Syrett, 'Corrosion monitoring using Harmonic Impedance Spectroscopy', in Proc. 12th Int. Corrosion Congress, NACE, 1993, 4200.
78. M. C. H. McKubre and B. C. Syrett, 'Harmonic Impedance Spectroscopy for the determination of corrosion rates in cathodically protected systems', in *Corrosion Monitoring in Industrial Plants Using Nondestructive Testing and Electrochemical Methods*, ASTM STP 908, ed. G.C. Moran and P. Labine. ASTM, Philadelphia, PA, 1986, 433.
79. N. G. Thompson and K. M. Lawson, *Development of an Advanced Corrosion Rate Monitor – The Harmonic Impedance Spectroscopy Method*. EPRI Report TR-107867, Electric Power Research Institute, Palo Alto, CA, 1997.
80. M. Kendig and D. Anderson, *Corrosion*, 48 (1992), 178.
81. H. Kihira, S. Ito and T. Murata, *Corrosion*, 45 (1989), 347.
82. L. Mészáros and J. Dévay, *Acta Chim. Acad. Sci. Hung.*, 105 (1980), 1.
83. R. W. Bosch and W. F. Bogaerts, *Corrosion*, 52 (1996), 204.

84. E. Kuş and F. Mansfeld, *Corros. Sci.*, 48(4) (2006), 965.
85. R. W. Bosch and W. F. Bogaerts, 'The small amplitude potential intermodulation technique for the monitoring of corrosion inhibition', Session II Corrosion Inhibitors, Proc. EUROCORR '96.
86. A. Rauf and W. F. Bogaerts, 'Critical pitting temperature can be measured using the EFM (Electrochemical Frequency Modulation) technique', Poster presentation, Corrosion 2005, NACE.

3

Electrochemical noise generated during stress corrosion cracking in high-temperature water and its relationship to repassivation kinetics

Rik-Wouter Bosch, Marc Vankeerberghen, Serguei Gavrilov and
Steven Van Dyck

SCK•CEN, Boeretang 200, B-2400 Mol, Belgium
rbosch@sckcen.be

3.1   Introduction

Electrochemical noise (EN) is a generic term used to describe spontaneous fluctuations of potential or current, which occur at the surface of freely corroding electrodes that can be picked up by a suitable electrode arrangement. Distinction is made between electrochemical potential noise and electrochemical current noise. Electrochemical potential noise is the fluctuation in the electrochemical potential of an electrode relative to a reference electrode, whereas electrochemical current noise is the fluctuation in an electrochemical current.

For the case of stress corrosion cracking (SCC), current fluctuations have been associated with the initiation, temporary propagation, and repassivation of intergranular cracks under boiling water reactor (BWR) conditions [1]. In this case, the initiations of significant cracks were shown as clear transient spikes, which could easily be discriminated from plots of the raw data. Under similar conditions (oxygenated, high-temperature water), potential noise measurements have been carried out with sensitised SS 304 O-rings. In this case, spikes in the potential time records were associated with crack initiation [2]. In addition, electrochemical noise measured under BWR conditions showed that different modes of cracking such as 'fan ribs', transgranular SCC, intergranular SCC and ductile fracture could be distinguished [3]. Under pressurised water reactor (PWR) conditions, the number of electrochemical noise tests is quite limited. Tests have been carried at 340°C with Inconel 600 during slow strain rate tests (SSRT). Both current and potential noise were measured. Analysis of the noise data revealed that both slip-dissolution and hydrogen embrittlement could be possible cracking mechanisms [4].

Electrochemical potential noise measurements have been carried out with sensitised stainless steel pressure tubes under PWR conditions [5]. Very short potential spikes, believed to be associated with crack initiation events, were detected when the sample was stressed above the yield strength. In this case, it was stated that potential noise had to be used, due to the low conductivity of the test solution. Current noise would then be mainly related to the IR-drop in the solution. Also the sample subjected to SCC should be completely electrically isolated from the chassis in the case of current noise, which would complicate the experimental set-up. On the other hand, it is general practice to try to measure both potential and current noise [6]. Current and potential noise signals are related to each other. Potential variations are caused by current variations (IR-drops) travelling through the electrolyte solution. In case of an environment with a high solution resistance, corrosion processes are very localised and in any case, difficult to detect with a noise sensor.

The slip-dissolution model of SCC states that crack-tip advance is based on active dissolution of metal atoms after film rupture and until full repassivation has occurred. The bulk water chemistry maintains a corrosion potential E_{corr} that will not change when a small crack due to film rupture is introduced. This bare metal surface will then repassivate as if the surface is polarised by E_{corr}. The peak current i_{active} will then exponentially decay to $i_{passive}$. This is illustrated in Fig. 3.1. The anodic current associated with the metal dissolution causes a potential drop across the electrolyte between the crack-tip and the place where this current is compensated by a cathodic reduction reaction.

If this potential 'noise' can be measured with a suitable electrode (for instance, a Pt wire sensor), the onset of SCC can be detected. A prerequisite for this is that part of the SCC is associated with repassivation, i.e. an anodic current causing a potential drop in the solution needs to be present. This is obvious for the slip-dissolution model, but other models might generate potential noise as well, albeit not in the rate

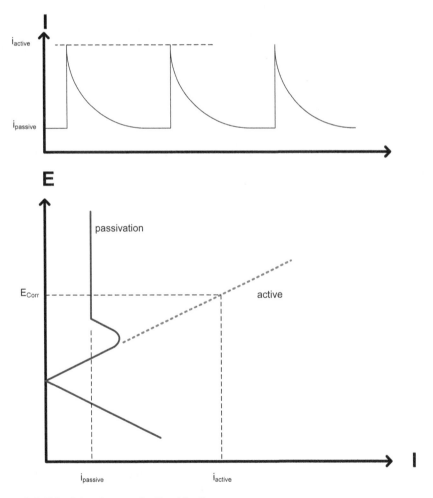

3.1 Principle of repassivation kinetics

determining step of the SCC model. For instance, hydrogen-induced cracking (HIC) and hydrogen-assisted cracking (HAC) are models that assume a dominant role of hydrogen in the cracking mechanism. In both models however, a bare metal surface is created after crack advance, which is repassivated before the next cracking event takes place.

Therefore, in this chapter, we discuss whether repassivation is the source of electrochemical noise (EN) during SCC and if electrochemical noise can be used to:

- Detect SCC
- Distinguish between initiation and propagation (?)
- Reveal the SCC mechanism (?).

For this purpose, results are shown from a scratch technique being a simulation for SCC, electrochemical potential noise measured during SCC and mathematical simulation of noise based on the repassivation equation. These results are then compared and we discuss whether this repassivation behaviour is indeed the source of SCC related electrochemical noise.

3.2   Repassivation with the scratch technique

The repassivation phenomenon can be investigated using a scratch test [7,8]. After a stainless steel sample is polarised in the passive part of the anodic polarisation curve, a sharp pin makes a scratch on the surface, hereby removing the passive film. Repassivation of the bare metal in the scratch will occur due to the applied potential. The current supplied to the metal sample is measured before, during and after scratching. A device has been designed that can perform such a test in an autoclave. A ceramic pin produces a scratch on a stainless steel sample when a metal bellows expands under a burst internal pressure. This principle is shown in Fig. 3.2.

The scratch unit consists of a metal sample connected to a metal bellows. This metal bellows moves the sample along three ceramic pins when the bellows is

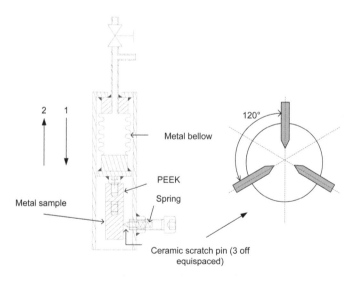

3.2 Schematic drawing of the scratch unit showing the metal sample, the ceramic scratch pins and their spring loading, the PEEK isolation part and the air-driven bellows which can move the sample up and down

expanding. The three ceramic pins are pushed through the oxide film on the sample by a spring. These springs are necessary to maintain a constant 'pressure' of the pins onto the metal surface. Simultaneously, the springs allow some flexibility, necessary to avoid fracture of the ceramic pins during scratching. The metallic sample is connected to the metal bellows by a PEEK (PolyEtherEtherKetone) transition piece to ensure electric isolation of the sample.

A typical test at high temperature was performed as follows. During heating of the autoclave, a bypass between the interior of the autoclave and the bellows maintained a pressure balance, preventing the bellows from expanding or contracting and making a scratch. After heating to the operating temperature, a stabilisation period was maintained for a minimum of 12 hours. Then an overpressure of a few bars was applied to move the bellows and produce a scratch on the sample surface. The polarisation curves and cyclic polarisation curves were measured on a separate sample in a range from -500 to $+1500$ mV with respect to the corrosion potential at a scan rate of 600 mV/h.

Here we will show some examples of repassivation currents measured after scratching of the oxidised metal surface. The first example was a test carried out at room temperature. Figure 3.3 shows the polarisation curve. From this curve, a passive potential was selected to polarise the stainless steel sample. Figure 3.4 shows the current-time plot. The potential has been fixed at $+100$ mV$_{SHE}$. Notice that the repassivation time is about 100 s, which seems slow. This relatively slow process is attributed to the measurement technique of scratching, where the measured current is a combination of repassivation, double layer charging and the finite speed of the scratching pin.

Figure 3.5 shows a typical result obtained in a solution of 400 ppm B and 2 ppm Li at 300°C in nitrogenated water (PWR primary water). A sharp increase in the current can be observed after which repassivation occurred and the current decreased. Whether this current can be solely related to repassivation kinetics has been the

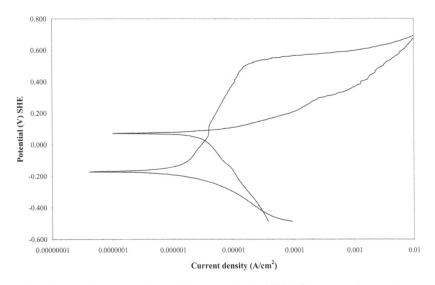

3.3 Polarisation curve for stainless steel in 0.5 M NaCl at room temperature

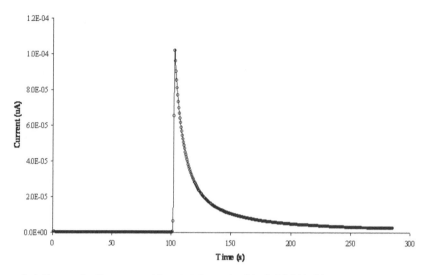

3.4 Repassivation current for stainless steel in 0.5 M NaCl at room temperature

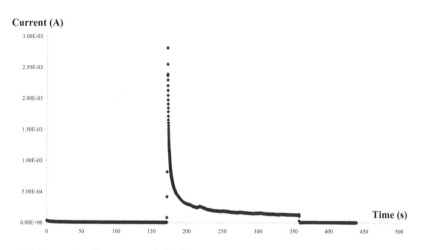

3.5 Repassivation current in high-temperature water

subject of some debate [9,10]. Due to the rapid rise in the current, the electrochemical double layer can contribute to the current as well. In addition, part of the current can be compensated by cathodic reactions at the sample surface and so is not measured at all, i.e. the current is not flowing via the counter electrode that measures the current.

The scratch technique is however a good simulation of the EN technique for SCC types of events. When the oxide layer at the crack-tip is ruptured, the anodic current will flow from the crack-tip to the solution. This current will be compensated by cathodic reactions at the crack wall and the outer surface of the crack. The potential drop caused by the latter reactions can eventually be picked up by a noise sensor, when placed at a suitable position.

3.3   Electrochemical noise during stress corrosion cracking

In this part, three examples of electrochemical noise data measured in different SCC systems are shown. These systems are

1. Cold worked stainless steel 304 subjected to a SSRT in oxygenated high-temperature water (280°C)
2. Sensitised stainless steel O-ring shaped samples subjected to constant load tests, but stressed at different levels [3]
3. A stainless steel 316 CW C(T) specimen tested under hydrogenated PWR conditions at 300°C.

3.3.1   Example 1

Figure 3.6 shows the first example of electrochemical potential noise measured in high-temperature water. When potential noise is measured during a cracking event, we would expect a sharp potential drop after which slow recovery of the potential will occur. During the test, only a few single crack events, as shown in Fig. 3.6, were observed. The fracture surface in Fig. 3.7 shows transgranular fracture. This is associated with the fact that the material was in the cold worked condition. With this type of SCC, the rate determining step is not necessarily related to repassivation kinetics.

3.3.2   Example 2

The noise pattern of Figs. 3.8 and 3.9 has been obtained with an O-ring compression test [2]. Potential noise was measured with a Pt wire of a stressed O-ring. The noise pattern does not reflect the 'ideal' shape of a repassivation curve, but could be related to the formation of many surface cracks as shown in Fig. 3.10.

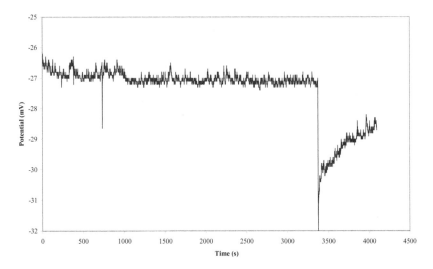

3.6 Noise event during SSRT of cold worked stainless steel in oxygenated high-temperature water

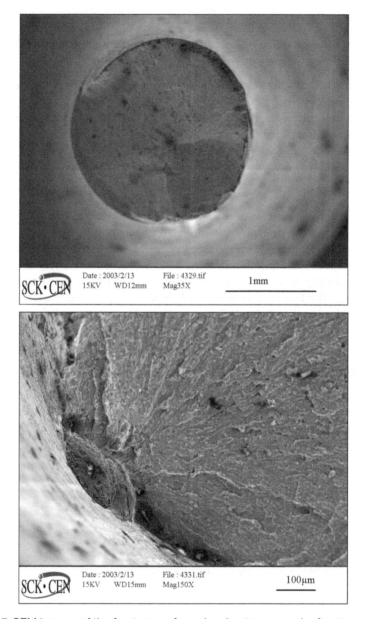

3.7 SEM images of the fracture surface showing transgranular fracture

3.3.3   Example 3

Figure 3.11 shows potential noise measured during a rising load test with a sub-size 316 CW C(T) specimen in PWR water (2 ppm Li, 400 ppm B, 2 ppm H_2, 300°C). The noise pattern indicates that a single crack event took place. Although the noise pattern is convincing, we might wonder whether it is possible to detect a single event. Analysis of the fracture surface after the tests showed little or no cracks related to SCC (Fig. 3.12). Some minor features of transgranular cracking can however be noticed.

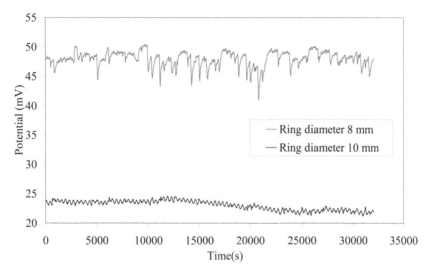

3.8 Multiple events related to SCC

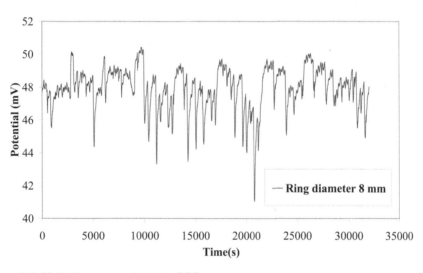

3.9 Multiple events related to SCC

3.4   Electrochemical noise generator based on repassivation kinetics

Standard deviation, skew and kurtosis are statistical parameters used to analysis noise data to identify certain processes such as pitting and SCC events. What would happen if we assume that a certain process will take place and, based on this, generate noise data? Would it be possible to analyse these 'ideal' noise data with these statistical parameters to retrieve the processes behind? Here we will try to do this by generating noise based on repassivation kinetics. The principle of this repassivation noise is illustrated in Fig. 3.13.

During repassivation, the current density is a strong function of the time elapsed since surface exposure. Generally, one assumes a power-law type function to express

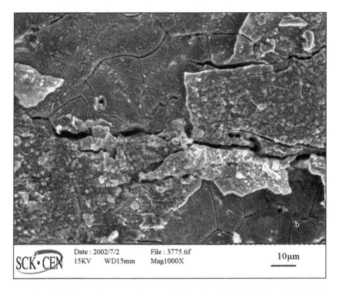

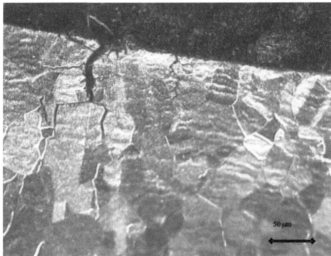

3.10 SEM images of the fracture surface showing transgranular fracture

this time dependency. Therefore, for the noise generator, the following equation has been used [7,8]

$$i(t_s) = i_{max} \left(\frac{t_0}{t_s + t_0} \right)^n + i_{pass},$$ [3.1]

where i is the current density, t_s is the time elapsed since the surface was exposed, i_{max} is the instantaneous current density increase at the moment of exposing the surface, i_{pass} is the passive current density and, t_0 and n are constants describing the repassivation kinetics. t_0 is usually associated with a time delay in the repassivation and n with the speed of the current decay under repassivation. Initially, the current density is equal to the sum of i_{max} and i_{pass} and at large elapsed time, it decays to i_{pass}.

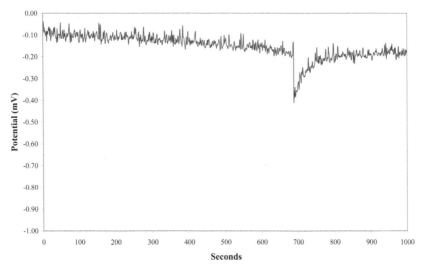

3.11 Single potential noise event related to SCC

$$\lim_{t_s \to 0} i(t_s) = i_{max} + i_{pass}$$

$$\lim_{t_s \to \infty} i(t_s) = i_{pass}$$

[3.2]

Another parameter used in the noise generator is the rupture frequency f. This value is not present in equation 3.1 but it determines how many rupture events occur within a certain time frame. So that means that we have three variables for the noise generator: the repassivation exponent n, the repassivation time constant t_0, and the rupture frequency f. In the following figures, the noise generated for $n=0.5, f=0.1$ and $t_0=0.001$ (Fig. 3.14) or $t_0=0.2$ (Fig. 3.15) is shown. Figure 3.14 still shows the single noise events related to repassivation currents. Figure 3.15 shows a noise pattern where single repassivation currents can no longer be distinguished. Notice that both noise patterns have been generated by the same repassivation current in equation 3.1.

This noise generator allows us to test different analysis methods to see if useful information could be retrieved from the noise data. Therefore we calculated the standard deviation, skew and kurtosis for each set of data and plotted these values as a function of n, t_0 and f. The result is shown in Fig. 3.16.

Figure 3.15 shows that different combinations of n and t_0 give different values of the three statistical parameters standard deviation, skew and kurtosis. For example, when we analysed the noise data with $n=0.5$ and $t_0=0.01$ we did not find any effect based on the skew and kurtosis, but did find a significant effect on the standard deviation. Alternatively, when we analysed the noise data with $n=1.5$ and $t_0=0.010$ it was the other way around. We did not find any effect based on the standard deviation, but did find a significant effect on the skew and kurtosis.

With this electrochemical noise simulator based on repassivation kinetics, similar noise patterns could be generated to those measured in high-temperature water. The noise simulator also demonstrated that statistical analysis of the noise could reveal the different repassivation parameters used. The scratch technique could be used to

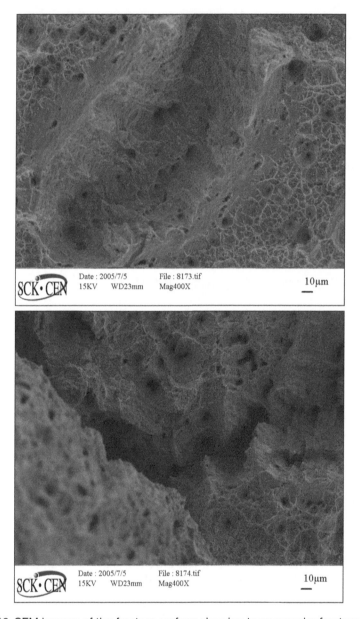

3.12 SEM images of the fracture surface showing transgranular fracture

experimentally demonstrate the principle of a single SCC event. This could be related to one of our experiments and therefore is useful. The scratch surface is however much larger than a real expanding crack. There are also some drawbacks with the technique used such as interference by electrochemical double layer charging, consumption of the anodic current by the passivated surface due to the low conductivity of the electrolyte solution and local change of the water chemistry due to the high local currents. An experiment where the full surface would undergo a film rupture would be more representative and is the topic of further investigations.

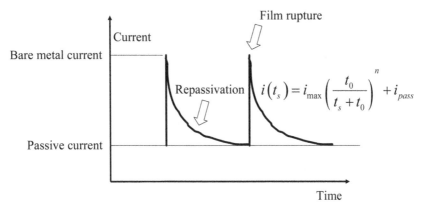

$$i(t_s) = i_{max} \left(\frac{t_0}{t_s + t_0} \right)^n + i_{pass}$$

3.13 Principle of the repassivation noise simulation model

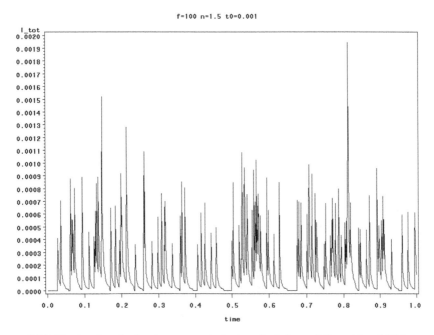

3.14 Electrochemical noise generated when $f = 0.1$, $n = 0.5$ and $t_0 = 0.001$

3.5   Conclusions

In this chapter, we tried to create a link between repassivation kinetics, electroche-mical noise and SCC. To measure an electrochemical noise signal that can be related to a cracking event, an anodic current from a bare metal surface is required. This anodic current does not necessarily have to be the rate determining step in the SCC process, but it has to be related to a cracking event. This anodic current causes a potential drop in the electrolyte solution that can be picked up by a suitable electrode.

Experimental simulation of this phenomenon can be carried out with a scratch test, which generates a single event current response. Some noise tests show similar

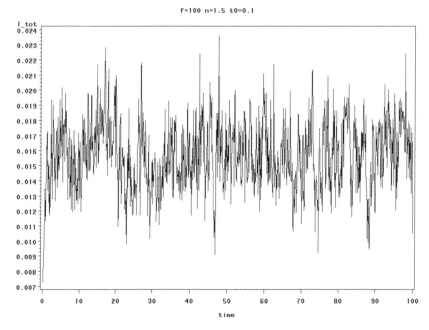

3.15 Electrochemical noise generated when $f = 0.1$, $n = 0.5$ and $t_0 = 0.2$

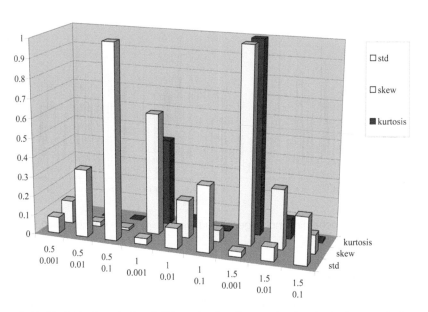

3.16 Statistical analysis of different sets of repassivation-based generated noise data; repassivation parameter n had values 0.5, 1.0, 1.5, $t_0 = 0.001$, 0.01 and 0.1, and the rupture frequency was $f = 0.1$

behaviour albeit that a single noise event hardly occurs. In most cases, the different repassivation peaks do overlap. Multiple events give a more realistic noise signal.

Simulations based on repassivation kinetics show how artificial noise patterns can be generated. Different noise patterns can be calculated based on the same repassivation equation. These data can be used to test different analysis methods. Here it was shown that standard deviation, skew and kurtosis do not give the same results and therefore have the potential to differentiate between various SCC mechanisms.

References

1. J. Stewart, D. B. Wells, P. M. Scott and D. E. Williams, *Corros. Sci.*, 33 (1992), 73.
2. S. Van Dyck and R. W. Bosch, 'Stress corrosion cracking initiation in sensitized stainless steel 316l exposed to oxygenated water', in Int. Conf. on Environmental Degradation of Materials in Nuclear Power Systems – Water Reactor, Stevenson, USA, 2003.
3. C. R. Arganis-Juarez, J. M. Malo and J. Uruchurtu, *Nucl. Eng. Des.*, 237 (2007), 2283.
4. Y. Watanabe and V. Kain, *JSME Int. J. Ser. A*, 45(4) (2002), 477.
5. R. Van Nieuwenhove, *Corrosion*, 56 (2000), 161.
6. R. Cottis and S. Turgoose, 'Electrochemical impedance and noise', in *Corrosion Testing Made Easy*, ed. B.C. Syrett, NACE International, 1999.
7. H. S. Kwon, E. A. Cho and K. A. Yeom, *Corrosion*, 56 (2000), 32.
8. R. W. Bosch, B. Schepers and M. Vankeerberghen, *Electrochim. Acta*, 49 (2004), 3029.
9. R. P. Wei and M. Gao, *J. Electrochem. Soc.*, 138 (1991), 2601.
10. P. D. Bastek, R. C. Newman and R. G. Kelly, *J. Electrochem. Soc.*, 140 (1993), 1884.

4

Detection of stress corrosion cracking in a simulated BWR environment by combined electrochemical potential noise and direct current potential drop measurements

Stefan Ritter and Hans-Peter Seifert

Paul Scherrer Institute (PSI), Laboratory for Nuclear Materials, Nuclear Energy and Safety Research Department, 5232 Villigen PSI, Switzerland

stefan.ritter@psi.ch

4.1 Introduction

In the chemical and power generation industry, corrosion-resistant, austenitic stainless steels are widely used as construction materials for piping, vessels and other structural components which enclose or come into contact with aggressive or mild environments and are subjected to relevant sustained mechanical loads. Intergranular stress corrosion cracking (SCC) in these materials is therefore a common phenomenon, in particular, if the material is in a susceptible condition (e.g. sensitisation, cold work, etc.). In the nuclear industry, intergranular SCC incidents have occurred in both boiling water (BWR) and pressurised water reactors (PWR) in a wide range of stainless steel and nickel-base alloy components such as reactor internals, (reactor) pressure vessel penetrations and nozzles, main coolant piping and heat exchanger tubing [1–3]. These SCC incidents have significantly affected plant availability/economics and, in some cases, have compromised the integrity of the primary circuit and thus plant safety.

The current pressure of competition in the power generation industry requires implementation of optimised, strategic and proactive plant ageing and life management methods for cost-effective operation of nuclear power plants at high safety levels. The ongoing occurrence of SCC in BWRs and PWRs worldwide clearly demonstrates a need for the development of advanced, non-destructive, continuous monitoring tools for the early detection of SCC initiation in the technical pre-crack stage.

The electrochemical noise (EN) measurement technique is a promising tool for continuous, in-situ corrosion monitoring in technical systems and has the potential to detect the nucleation and initiation of localised corrosion processes [4]. Currently, EN seems to be the major promising technique with some potential for the early detection of SCC initiation in structural components of BWRs and PWRs, although no systematic studies have been performed so far [5,6]. However, the basic and quantitative understanding of EN during SCC is rather limited.

Therefore, the feasibility of the EN technique as a corrosion monitoring tool under BWR conditions is being investigated at PSI in the framework of the KORA project

[7]. The experiments performed so far clearly indicate that early detection of SCC initiation by EN might be possible in a simulated BWR/normal water chemistry environment under lab conditions and that single, small, semi-elliptical surface flaws with a crack length and depth in the range of about 150 μm can be detected by EN measurements [7,8]. A big challenge in the evaluation of this monitoring technique under high-temperature water conditions is to find an independent SCC detection method that could be applied simultaneously to verify the few, but very promising, results of EN investigations performed earlier at PSI [7] and elsewhere [9–12]. The direct current potential drop (DCPD) method, which has been used for online crack growth measurements in compact tension (C(T)) specimens for many years, might be such an independent SCC monitoring tool. In the current paper, the most important results from combined EN[1] and DCPD measurements are briefly summarised.

4.2   Material and experimental procedure

4.2.1   Material and specimens

For the SCC initiation tests, a rod material of the high-carbon austenitic stainless steel AISI 304 was used, since this could be easily sensitised and showed a sufficiently high susceptibility to intergranular SCC (Tables 4.1 and 4.2). This material was used either in the solution annealed state with a rather low degree of sensitisation (value from double-loop electrochemical potentiokinetic reactivation (DL-EPR) tests ≤0.3%; measured according to JIS G 0580: 1986) and therefore with a low susceptibility to intergranular SCC, or it was sensitised. Heat treatment at 620°C for 24 h resulted in a very high degree of sensitisation (DL-EPR value ≈ 28%). In the tests, 12.5 mm thick compact tension (0.5 T C(T)) specimens with a machined notch ($\rho = 0.1$ mm) were used (Fig. 4.1a). Two specimens were tested simultaneously in a 'daisy-chain' arrangement. Additional tests were performed using round bar specimens with a U-shaped notch in the centre of the gauge section (Fig. 4.1b, gauge length = 36 mm, diameter = 6 mm, notch depth = 0.2 mm). All specimens were electrically insulated from the autoclave and from each other and were instrumented for EN and DCPD techniques (see below).

4.2.2   Experimental procedure

Sophisticated, refreshed, high-temperature water loops with autoclaves and integrated mechanical loading systems were used for the SCC initiation studies with simultaneous EN and DCPD measurements under simulated BWR conditions (Fig. 4.2). This system was optimised for these kinds of measurement in many pretests. Additionally, the EN measurement devices were qualified and characterised by suitable electronic circuits and 'round robin' tests (e.g. within the European Cooperative Group on Corrosion Monitoring of Nuclear Materials – ECG-COMON) [13].

The SCC initiation tests were performed in oxygenated, high-temperature water (250°C or 288°C with 2 ppm dissolved oxygen (DO) content). The electrochemical

---

[1] The term electrochemical potential *noise* (EPN) as used in this paper includes more 'normal' measurement of electrochemical potential.

Table 4.1 Mechanical properties and heat treatments of the investigated material

AISI	Design.	Type	Product form	YS$_{air}^{25°C}$ [MPa]	UTS$_{air}^{25°C}$ [MPa]	YS$_{air}^{288°C}$ [MPa]	Solution annealing heat treatment	Sensitisation heat treatment
304	304C	High-carbon	Rod	291	686	175	1050°C/30 min/WQ	–
304	304D	High-carbon	Rod	291	686	175	1050°C/30 min/WQ	620°C/24 h/WQ

WQ, water quenched.

Table 4.2 Chemical composition of the investigated material in wt.-%

AISI	Design.	C	Si	Mn	P	S	Cr	Mo	Ni	Co	Cu	Nb	Ti
304	304C&D	0.062	0.15	0.53	0.023	0.029	18.3	0.273	8.59	0.169	0.296	0.019	0.001

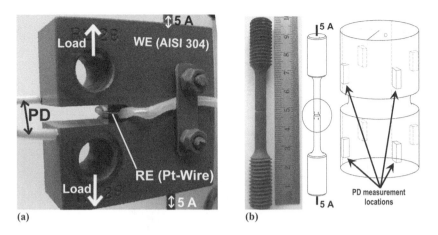

4.1 12.5 mm thick compact tension specimen (0.5 T C(T)) with machined notch and instrumented with direct current potential drop (DCPD) and electrochemical potential noise measurement technique (a); round bar specimen with U-shaped notch and schematic of the potential drop (PD) measurement locations (b)

corrosion potential (ECP) of the specimens and the redox potential of the environment (Pt-probe) were measured (during the pre-oxidation phase only) with a Cu/Cu$_2$O/ZrO$_2$-membrane reference electrode. The ECP of the stainless steel was about $110_{(250°C)}$ or $160_{(288°C)}$ mV$_{SHE}$[2] and the redox potential was $270_{(250°C)}$ or $300_{(288°C)}$ mV$_{SHE}$. In most experiments, 50 ppb sulphate was added to the high-purity water to promote SCC in the stainless steel specimens. The specimens were pre-oxidised at a small pre-load for at least 1 week before constant pull-rod stroke rates ($v_{pull\,rod}$) of $2·10^{-8}$ or $1·10^{-8}$ m/s (C(T) specimens) or $7.2·10^{-9}$ m/s (round bar specimens) were applied. After SCC initiation and a certain amount of crack advance was detected by DCPD and/or EN measurement, the control mode was switched from constant extension rate testing (CERT) to load control. The specimens were then either directly unloaded, or the load was kept constant until unloading at the end of the test. After finishing the experiments, the specimens were opened by fatigue (C(T) specimens) or by mechanical overloading (round bar specimens) in air at room temperature for analysis of surface/in-depth cracking and of the fracture mode (inter- or transgranular SCC, ductile mechanical failure, etc.) using a scanning electron microscope.

Electrochemical noise measurements

The electrochemical potential noise (EPN) was recorded either by a GAMRY™ potentiostat (ESA400 software) with a sampling rate of 5 Hz, or by a newly developed EN measurement device (EcmNoise, IPS) with a sampling rate of 2 Hz. The EPN was measured versus a Pt-wire (pseudo reference electrode) which was inserted into the notch of the C(T) specimen or aligned around the U-notch of a round bar specimen and located as close as possible to the notch-root (Figs. 4.1a and 4.3a) or U-notch, respectively. Due to simultaneous use of the EN and DCPD techniques, the electrochemical current noise (ECN) could not be measured. Despite a

---

[2] The ECP was usually still slowly increasing at the end of the pre-oxidation phase.

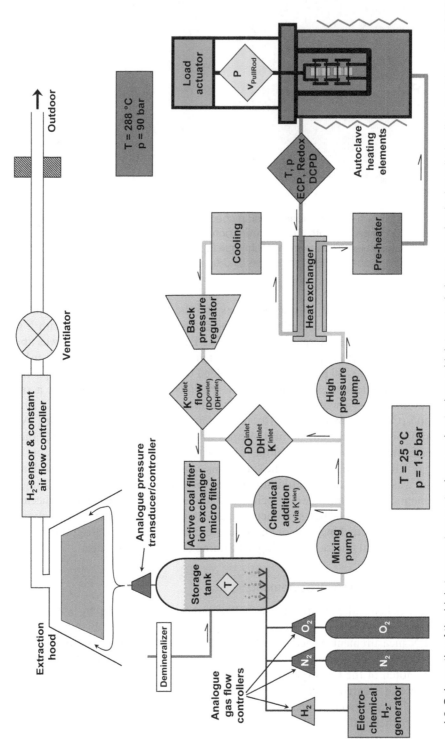

4.2 Schematic of the high-temperature water loop and autoclave with integrated electro-mechanical tensile loading machine with DCPD system for the SCC initiation studies under simulated BWR conditions

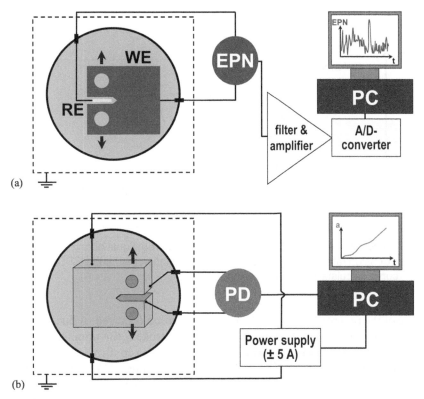

4.3 Schematic of the cell configuration and experimental set-up for electro-chemical potential noise (EPN) (a) and reversed direct current potential drop measurements (b) with C(T) specimens. The two techniques were combined in an autoclave

pre-oxidation phase of more than 1 week before starting the loading phase, the EPN signal was still drifting slowly; quasi-stationary ECPs and passive current density values on stainless steels are usually achieved only after several thousands of hours of exposure in high-temperature water. This slow change in ECP from oxide film growth has to be carefully considered in EN analysis. Therefore, the DC part and the trend of the EPN signal were removed for evaluation of the data, resulting in a basic EPN level of around 0 mV when the loading phase started ($t = 0$ h).

Direct current potential drop measurements

Together with the EPN, crack advance/initiation was continuously monitored using the reversed DCPD method with a resolution limit of about 1 µm. Figure 4.3b shows a schematic of the DCPD measurement set-up for the C(T) specimens. A current of 5 A was sent through the specimen (the polarity was reversed about every 3 s) and the potential drop was measured at the front face of the specimen. In the case of the round bar specimen, a current of 5 A was also used and the potential drop was measured at four locations around the U-notch (Fig. 4.1b).

The geometry change by plastic deformation of the round bar specimens and of the base of the notch in the C(T) specimens during the CERT phase resulted in a

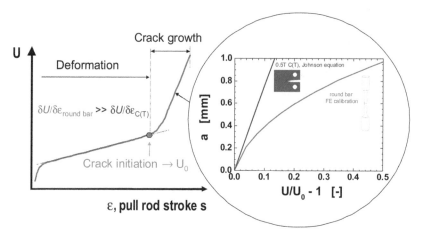

4.4 Schematic of the course of the potential drop signal during constant extension rate testing with point of crack initiation (left) and DCPD calibration curve calculated by FE modelling (right)

change of the potential drop U before crack initiation occurred. A linear relationship between change in U and increase in pull rod stroke s (or nominal strain ε) was observed during this phase, and the behaviour was also confirmed by Finite Element (FE) modelling (Fig. 4.4). The potential drop change $\delta U/\delta s$ was much higher (typically by a factor of 10) in the round bar specimens with plastic yielding of the whole cross-section (SCC initiation was typically observed at nominal stresses slightly above the yield strength) than in the notched C(T) specimen, where plastic yielding was restricted to the base of the notch (small scale yielding). The SCC initiation point (and potential drop at crack initiation U_0) during the CERT phase was assigned to the location where the U vs. s (or ε) curve started to deviate from the abovementioned linear relationship. The subsequent SCC crack growth was then calculated from U/U_0 and the analytical Johnson equation [14] for the case of C(T) specimens and by using our own calibration curve, which was derived by FE modelling (Fig. 4.4), for the round bar specimens. Although the DCPD has a higher sensitivity for crack growth $\delta U/\delta a$ in the round bar than in the C(T) specimens, exact determination of the point of crack initiation was subject to a larger degree of uncertainty because of the more pronounced plasticity effects.

4.3   Results and discussion

The first combined EN and DCPD measurements in high-temperature water revealed some interference to the EPN signal from the DCPD technique. Figure 4.5 shows the disturbances in the EPN signal that disappeared when the DCPD measurement was turned off. The steps in the EPN signal appeared synchronously with reversing the polarity of the current. This effect was probably caused by imperfect electrical insulation of the specimens and wires leading to very small leakage currents, which resulted in relevant ohmic potential drops in the low-conductivity electrolyte. Unfortunately, the experimental set-up could not be further improved and periodical interruptions of the DCPD measurements would have led to a rather poor resolution limit for the detection of SCC initiation. Therefore, some information in the EN signal was

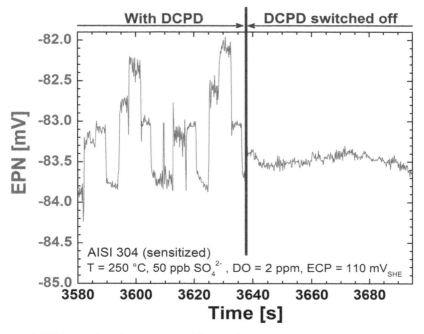

4.5 EPN signal during the pre-oxidation phase of the first experiment with reversed DCPD turned on (left) and off (right). An interference of the DCPD to the EPN signal could be identified

lost due to these disturbances. On the other hand, the influence on the EPN signal was still at an acceptable level and it was decided to continue with these tests and to concentrate on the mean value of the EPN signal by averaging the EPN values over periods of 5 min.

Individual potential (and current) transients, as they were observed during SCC initiation tests in a similar high-temperature water environment with notched round bar and flat tensile specimens without DCPD (Figs. 4.6 and 4.7), could therefore not be resolved, but a 'global' change in potential was still visible. Such a 'global' decrease in the potential (and increase in the current) always appeared if SCC initiation occurred (Figs. 4.6 and 4.7a) [7,15]. The observed cathodic (or anodic) polarity and the shape of individual potential (and current) transients are probably related to oxide film rupture and repassivation events during crack initiation at different surface locations, as well as to surface crack growth of previously formed microcracks. The initiation process of intergranular SCC thus involves local anodic dissolution, which would be in line with a slip dissolution mechanism [16,17]. The superposition of such potential (and current) signals from initiation events at different surface locations and the surface crack growth of microcracks under slow straining conditions with increasing plastic strain may result in a quasi-continuous drop in potential (and increase in the anodic current). Furthermore, superimposed crevice currents and resulting potential changes to the differential aeration cell in the crack-mouth region, which vary with the crack-mouth opening, further contribute to these signal changes. Individual transients therefore cannot be resolved in every case, in particular, for large distances between specimen surface and reference (or counter) electrode. In

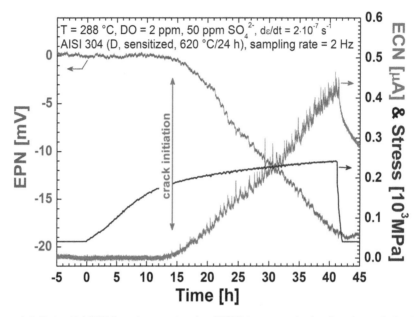

4.6 Potential (EPN) and current noise (ECN) in a constant extension rate test with a sensitised stainless steel round bar specimen with U-notch. The decreasing EPN and increasing ECN signal as well as some transients indicate SCC initiation

Fig. 4.8, another result of an earlier experiment with a round bar specimen is shown, where the specimen was unloaded very briefly (3.5 h) after a negative change in the EPN signal (and increase in the ECN signal) was observed, thus indicating SCC initiation. Post-test analysis with the scanning electron microscope in fact revealed one single, small, intergranular, semi-elliptical flaw with a surface crack length of about 150 μm and a maximum crack depth of about 100 μm (Fig. 4.8b).

Figure 4.9 shows the result of the first SCC initiation test with combined EN and DCPD measurements using a heavily sensitised stainless steel C(T) specimen. After about 8 h of constant straining of the specimen, the EPN signal started to drop indicating SCC initiation. This was confirmed by the DCPD technique, which showed the onset of crack advance at the same time. Subsequent fractographic analysis by scanning electron microscope revealed intergranular SCC initiation and growth (along the whole notch-root) in this specimen (Fig. 4.9b). Due to the disturbances of the EPN signal, the expected individual potential transients in the EPN from single local film rupture/repassivation events related to SCC initiation could not be resolved (see above). Investigations in aqueous solutions at room temperature, where a comparable cracking mechanism is believed to be active, revealed a similar behaviour of the EPN signal level if extensive SCC initiation occurred [16,17]. About 3 h after SCC initiation was detected, the loading mode was switched to constant load and the crack continued to grow according to the DCPD signal (Fig. 4.9a). Rather surprisingly, the EPN signal rose back close to the original level. The reason for this is probably the absence of any further surface crack growth along the notch-root and the high-purity water electrolyte with a very low conductivity (κ_{inlet} (at 25°C) $= 0.19$ μS/cm). Once the SCC surface crack had spread out along the whole

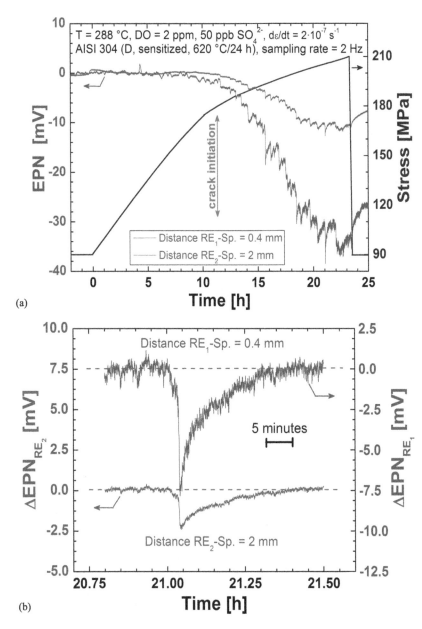

4.7 Effect of distance between reference electrode and specimen surface on the EPN during a constant extension rate test with a flat tensile specimen and two pseudo reference electrodes (Pt-wires) (a). Example of an individual potential transient from (a) at higher resolution (b)

notch-root, the active crack-tip apparently grew in the depth direction and moved away from the surface. Thus the EN signal from the crack front accessible outside the crack became small, in good agreement with the limited throwing power of the current in the high-purity water. Therefore, only crack initiation and subsequent surface (or near-surface) crack growth appear to be detectable by EN measurements

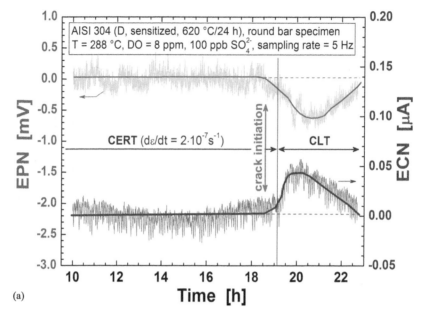

(a)

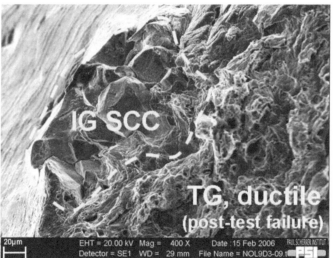

(b)

4.8 Potential (EPN) and current noise (ECN) in a combined constant extension rate (CERT)–constant load (CLT) test with a round bar specimen (sensitised stainless steel), which was switched to constant load in the range of the yield strength when the potential and current signals started to change and indicated SCC initiation (a). The specimen was unloaded after a further 3.5 h. Post-test analysis with the scanning electron microscope revealed one single, small intergranular semi-elliptical surface flaw (b)

in low-conductivity electrolytes. Recent, preliminary results from experiments in high-purity water with multiple reference electrodes investigating the effect of reference electrode distance to the specimen surface (or 'location of cracking') seem to confirm this hypothesis (Fig. 4.7). After SCC initiation, the drop in the EPN signal

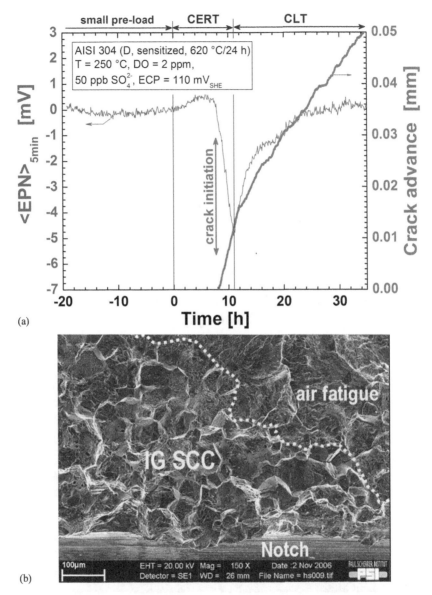

(a)

(b)

4.9 Mean values of the electrochemical potential noise (<EPN>) and crack advance measured by DCPD in a combined constant extension rate (CERT)–constant load (CLT) test with a sharply notched, sensitised stainless steel C(T) specimen (a) with intergranular SCC confirmed by post-test fractography with the scanning electron microscope (b)

(Fig. 4.7a) and single potential transients (Fig. 4.7b) could increasingly be resolved with decreasing distance to the specimen surface.

A second SCC initiation test was performed with sensitised (Fig. 4.10) and solution annealed (Fig. 4.11) stainless steel C(T) specimens at 288°C and with a slower extension rate of the pull rod ($v_{pull\ rod} = 10^{-8}$ m/s). This time, the test was interrupted shortly after switching to constant load. SCC initiation was detected in both specimens

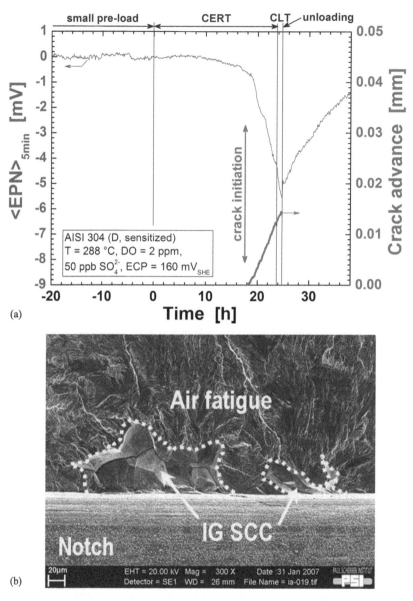

(a)

(b)

4.10 Mean values of the electrochemical potential noise (<EPN>) and crack advance measured by DCPD in a combined constant extension rate (CERT)–constant load (CLT) test with a sharply notched, sensitised stainless steel C(T) specimen (a) with localised, intergranular SCC confirmed by post-test fractography with the scanning electron microscope (b)

around 18 h after the start of the straining by a drop in the EPN signal and by crack advance recorded by the DCPD technique. In the sensitised stainless steel specimen, ca. 15 μm average crack advance over the whole specimen thickness was measured by the DCPD and the EPN signal decreased by ca. 5.5 mV, whereas in the solution annealed specimen, a total crack advance of ca. 9 μm and a decrease in the EPN by

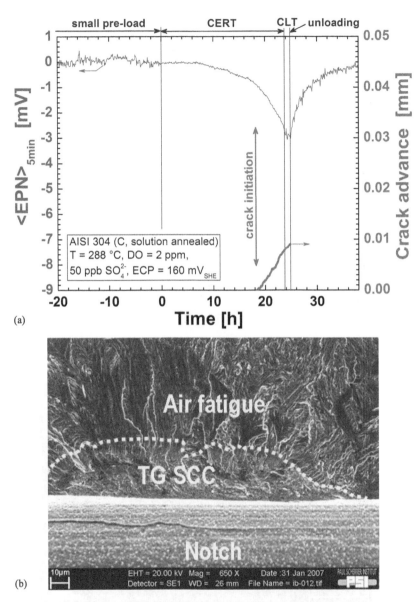

(a)

(b)

4.11 Mean values of the electrochemical potential noise (<EPN>) and crack advance measured by DCPD in a combined constant extension rate (CERT)–constant load (CLT) test with a sharply notched, solution annealed stainless steel C(T) specimen (a) with very localised, transgranular SCC confirmed by post-test fractography with the scanning electron microscope (b)

ca. 3 mV were recorded. In both specimens, surface crack growth along the notch-root could be followed by the EN measurements. Post-test fractography revealed a few, localised, intergranular areas of SCC in the sensitised specimen (Fig. 4.10b) and very few, localised, transgranular areas of SCC in the solution annealed specimen (Fig. 4.11b). The average extent of crack advance correlated reasonably well with the DCPD indications.

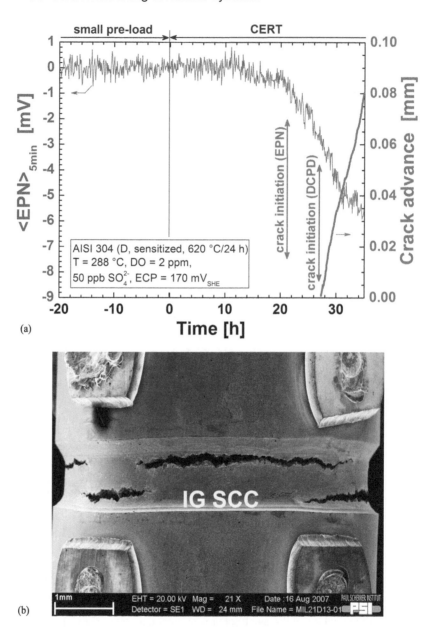

(a)

(b)

4.12 Mean values of the electrochemical potential noise (<EPN>) and crack advance measured by DCPD during constant extension rate testing (CERT) with a notched, sensitised stainless steel round bar specimen (a) with intergranular SCC confirmed by post-test fractography with the scanning electron microscope (b)

This second test also showed that similar EN signals are observed for intergranular and transgranular SCC in sensitised and solution annealed stainless steels in high-temperature water, indicating that the cracking mechanism involves similar electrochemical processes. Thus, differentiation between inter- and transgranular SCC will hardly be possible based on EN measurements alone.

To further verify the very promising EN results gathered during many SCC initiation tests with round bar specimens (e.g. Figs. 4.6 and 4.8), the EN and DCPD techniques were also applied simultaneously using notched round bar specimens. The result of one of these tests is shown in Fig. 4.12. After about 20 h of constant straining, the typical drop in the EPN signal was observed, whereas the DCPD signal indicated the onset of SCC several hours later. The delayed indication of SCC might be explained by the more pronounced effect of plastic deformation on the DCPD signal in the round bar specimen (see section 4.2.2.2) and by a greater sensitivity to small cracks for the EN technique.

Additional SCC initiation tests with combined EN and DCPD measurements on C(T) and round bar specimens further confirmed the observations described above [8,15]. These SCC initiation experiments with independent on-line crack growth monitoring by the DCPD technique verified the general ability of the EN technique to detect SCC initiation under simulated BWR conditions. The detection sensitivity increases with decreasing distance between specimen surface and reference (or counter) electrode [18]. Therefore, it is concluded that with suitable positioning of the reference (or counter) electrode, early SCC detection by the EN technique is possible under stable and stationary laboratory conditions in oxygenated high-temperature water.

4.4   Summary and conclusions

A big challenge in evaluation of the EN method as a SCC monitoring tool under high-temperature water conditions is to find an independent SCC detection method which can be applied simultaneously to verify promising results obtained in earlier investigations at PSI [7] and elsewhere [9–12]. Experiments with sharply notched compact tension specimens and U-shape notched round bar specimens showed that the reversed DCPD technique can be applied together with EPN measurements in an oxygenated high-temperature water environment as a second independent method to detect SCC initiation. In all investigated specimens, SCC initiation could be detected at around the same time by both techniques. Therefore it is concluded that early SCC detection by EN is possible under simulated BWR conditions, at least under stable and stationary laboratory conditions. However, only crack initiation and subsequent surface (or near-surface) crack growth can be detected by EN measurements in high-purity water with very low conductivity and a short distance between the specimen surface and the reference (or counter) electrode is crucial to achieve high sensitivity here. The observed polarity of the potential signal changes/transients during SCC initiation and growth suggests a mechanism which involves film rupture and local anodic dissolution/repassivation. Further investigations are still necessary to confirm these preliminary results.

Acknowledgement

The results described in this paper were generated within the KORA research programme. The financial support of the programme by the Swiss Federal Nuclear Safety Inspectorate (HSK) is gratefully acknowledged. Thanks are also expressed to B. Baumgartner, L. Nue, U. Tschanz, B. Gerodetti, and E. Groth[†] (all PSI) for their experimental contribution to this work and to K. Reichlin (PSI) for the FE calculations.

References

1. H. P. Seifert, S. Ritter and J. Hickling, *PowerPlant Chem.*, 6(2) (2004), 111–123.
2. R. Kilian and A. Roth, *Mater. Corros.*, 53 (2002), 727–739.
3. P. Scott, 'An overview of materials degradation by stress corrosion cracking in PWRs', in EUROCORR 2004, EFC, Paper No. 563 (Nice, France, 12–16 September 2004).
4. S. Ritter, T. Dorsch, and R. Kilian, *Mater. Corros.*, 55(10) (2004), 781–786.
5. J. Göllner, *Mater. Corros.*, 55(10) (2004).
6. S. Ritter and H. P. Seifert, *Detection of Stress Corrosion Cracking by the Electrochemical Noise Technique–State-of-the-Art*, TM, TM-43-07-01, Paul Scherrer Institute, Villigen PSI, Switzerland, January 2007.
7. H. P. Seifert, et al., *KORA: Environmentally-Assisted Cracking in Austenitic LWR Structural Material*, HSK Research Report, Paul Scherrer Institute, Villigen, Switzerland, November 2006.
8. H. P. Seifert, S. Ritter, B. Baumgartner and L. Nue, *KORA: Environmentally-Assisted Cracking in Austenitic LWR Structural Material*, HSK Research Report, Paul Scherrer Institute, Villigen, Switzerland, November 2007.
9. J. Hickling, D. F. Taylor and P. L. Andresen, *Mater. Corros.*, 49(9) (1998), 651–658.
10. Y. Watanabe, T. Shoji and T. Kondo, 'Electrochemical noise characteristics of IGSCC in stainless steels in pressurized high-temperature water', in NACE Corrosion, NACE, Paper No. 129 (San Diego, CA, USA, 22–27 March 1998).
11. J. Stewart, D. B. Wells, P. M. Scott and D. E. Williams, *Corros. Sci.*, 33(1) (1992), 73–88.
12. S. van Dyck and R.-W. Bosch, 'A study of stress corrosion crack initiation in high-temperature water using electrochemical noise on compressed ring samples', in 2nd Int. Conference on Environmental Degradation of Engineering Materials 2003, EFC (Bordeaux, France, 29 June–2 July 2003).
13. Website of the European Cooperative Group on Corrosion Monitoring of Nuclear Materials, available from http://www.ecg-comon.org (viewed on 1 April 2010)
14. H. P. Seifert and S. Ritter, *PSI Contribution to the CASTOC Round Robin on EAC of Low-Alloy RPV Steels under BWR Conditions*, 01–08, PSI-Report, Paul Scherrer Institut, Villigen, August 2001.
15. S. Ritter and H. P. Seifert, 'Combination of EPN and DCPD measurements under simulated BWR conditions – Preliminary results', in Annual Meeting of the European Cooperative Group on Corrosion Monitoring of Nuclear Materials, Paper No. WG2-4 (Magdeburg, Germany, 18–19 June 2007).
16. S. Ritter and H. P. Seifert, 'Detection of SCC initiation by EN measurements in thiosulphate solution at room temperature', in Annual Meeting of the European Cooperative Group on Corrosion Monitoring of Nuclear Materials, Paper No. WG1-5 (Villigen PSI, Switzerland, 16–17 June 2008).
17. Y. Watanabe and T. Kondo, 'Current and potential fluctuation characteristics in IGSCC processes of stainless steels', in NACE Corrosion, NACE, Paper No. 376 (San Diego, CA, USA, 22–27 March 1998).
18. H. P. Seifert, et al., *KORA: Environmentally-Assisted Cracking in Austenitic LWR Structural Material*, HSK Research Report, Paul Scherrer Institute, Villigen, Switzerland, November 2008.

5

Electrochemical noise measurements of stainless steel in high-temperature water*

Carlos R. Arganis-Juarez

Instituto Nacional de Investigaciones Nucleares, Km. 36.5, Carretera Federal México-Toluca, Municipio de Ocoyoacac, C.P. 52045, Estado de México, México

José M. Malo

Instituto de Investigaciones Electricas, Av. Reforma 113, Col. Palmira, C.P. 62490, Cuernavaca, Morelos, México

jmmalo@iie.org.mx

J. Uruchurtu

Centro de Investigaciones en Ingeniería y Ciencias Aplicadas, Universidad Autónoma del Estado de Morelos, Av. Universidad 1001, Col. Chamilpa, C.P. 62210, Cuernavaca, Morelos, México

5.1 Introduction

Until the 1970s, many of the corrosion problems in nuclear boiling water reactors (BWRs) were related to stress corrosion cracking (SCC), particularly to intergranular SCC (IGSCC). This was explained by the fact that austenitic steels undergo chromium carbide precipitation at grain boundaries when heated in the 600–700°C interval for a certain period. This phenomenon is known as sensitisation and such precipitation produces chromium depleted zones next to the precipitates, leading to a higher corrosion rate at grain boundaries compared to the steel matrix. This, along with the residual and operational stresses and impurities of the demineralised water used in this type of reactor at 288°C and 8 MPa pressure, contribute to the IGSCC phenomenon.

Even when a number of measures were taken to mitigate this problem, new forms of cracking have emerged, particularly in the internal components of the reactor pressure vessel. Although these components are built from a low carbon austenitic 304L steel in which short-term sensitisation is not possible, the fluence of fast neutrons above 1 MeV produces radiation induced segregation (RIS), which in turn leads to chromium depletion at grain boundaries. Analyses have also predicted high residual stress levels, particularly in the core shroud welds. The magnitude can exceed the material's annealed yield strength, in particular, if the surface has any cold

* Reprinted from *Nuclear Engineering and Design*, 237(24), C.R. Arganis-Juarez, J.M. Malo and J. Uruchurtu, 'Electrochemical noise measurements of stainless steel in high temperature water', pp. 2283–91, © 2007, with permission from Elsevier.

working before final joining. While these factors are important contributors, the very high oxidising nature of the core coolant at all elevations also appears to be a very strong driver for the cracking of the generally resistant L-grade or stabilised stainless steels [1].

In view of the numerous studies on the cracking rate that have been carried out on thermally sensitised materials, an equation has been proposed [2] which correlates the 'equivalent' thermal sensitisation, corresponding to the same rate of cracking, with the chromium depletion produced by irradiation plus any initial thermal sensitisation of the component before exposure to the neutron fluence. Therefore the study of cracking of thermally sensitised austenitic stainless steels is still useful because it can be related to the 'equivalent' fluence (neutron flux · time) that produces the same sensitisation.

The electrochemical noise (EN) technique has been used in the nuclear industry to detect pitting and crevice corrosion in high conductivity media, such as the service water of heat exchangers and steam generators of pressurised water reactors (PWRs) [3]. However, in BWRs, the extremely low conductivity of demineralised water, as well as the high temperature and pressure, is a problem for the extended use of electrochemical methods. This limits the choice of materials for construction of electrodes and sensors and consequently available data in the literature of this type are rather limited.

In previous works with slow strain rate tests (SSRTs) [4,5] in high oxygen and high-temperature water environments and low conductivities, EN current signals from 0.5 to 3 µA with very short rise time from 2 to 60 s and slow decays of 100 to 1000 s were observed. However, the authors did not measure potential noise signals. On the other hand, Andresen et al. [6] worked with potential signals and used voltage standard deviation but did not use current signals. Apparent strain rates of 7.9E-6, 1E-6 and 3.3E-7 s^{-1} were used in these investigations.

Wang et al. [7] worked in similar environments with C-ring specimens and obtained current and voltage signals. The current signals were about 1 µA, but the signals were related to changes in the stress intensity factor K and do not present typical transients caused by SCC. Finally Macdonald et al. [3] obtained periodic bursts of higher frequency noise of 100 µA amplitude with compact tension specimens at 33 MPa m$^{1/2}$ with a platinised nickel cathode, also related to changes in K.

In all these investigations, the mechanism of IGSCC was related to transients in current, but only a few measured electrochemical potential noise, and transgranular SCC (TGSCC) or the different sub-type fracture modes, reported in other studies with SSRTs [8] in stainless steel in demineralised water at 288°C, were not mentioned in these studies.

Due to its online monitoring capabilities, the EN technique has been considered by the nuclear industry [2] to be potentially part of smart systems that involve the use of predictive cracking models, water radiolysis, electrochemical potential measurements, and the use of sensors and codes to survey the structural integrity of BWRs.

5.2   Experimental procedure

5.2.1   Materials

The 304 steel samples were obtained from a 25.4 mm plate that underwent solution annealing heat treatment (1 h at 1050°C and water-cooled) to reduce the presence of

Table 5.1   Composition of the investigated steel (in wt.%)

Steel	Cr	Ni	C	Mn	Si	P	S.
304	18.39	10.45	0.046	1.65	0.41	0.011	0.001

Table 5.2   Characterisation of the investigated steel

Heat treatment	EPR or Pa, coulombs/cm^2	I_r/I_a	A-262*
304 Solution annealed. (1050°C 1 h + WQ)	0.3547	0.002	Step
304 Sensitised (1050°C 1 h + WQ + 650°C 4 h + WQ)	26	0.666	Ditch

*ASTM A-262 Standard Practice A Oxalic Acid Etch Test Microstructures. All present delta ferrite ribbons.
WQ, water quenching.

carbide sensitisation at the grain boundaries as much as possible. Some samples underwent a sensitisation heat treatment (4 h at 650°C and water-cooled) to promote the precipitation of carbides at the grain boundaries.

Table 5.1 shows the chemical composition of the investigated steel, and Table 5.2 shows the sensitisation parameters for the steel with the heat treatments.

5.2.2   Slow strain rate tests

Stress probes were prepared from 304 plate, 30 mm long (gauge section) and 3.5 mm diameter. Due to possible problems with cracking remote from the sensors, a small 0.5 mm deep notch was machined in the probes with a 60° angle. The SSRTs were carried out in a loop, comprising a clean-up and water conditioning unit and an autoclave with integrated mechanical test machine, as shown in Fig. 5.1. The inlet conductivities were 0.06 µS/cm in all cases and outlet conductivities were also reported. The SSRTs were conducted at a displacement rate of 3E-5 mm/s which resulted in an 'apparent strain rate' of 1E-6 s^{-1}.

Previous calibration was performed with sensitised stainless steel and it was found that the water saturated with atmospheric air resulted in a dissolved oxygen content of about 4000 ppb at 25°C (3000 m above sea level) and produced an electrochemical potential around 200 mV$_{SHE}$. The measurements were made with an external Ag/AgCl reference electrode. This condition was selected for the experiments which simulated the environment in the core of a BWR with values near 200 mV$_{SHE}$ due to the water radiolysis products.

5.2.3   Electrochemical noise instrumentation

The arrangement for monitoring EN, designed to insulate the working electrode from external interference, is shown in Fig. 5.2. Two small area platinum electrodes were used as reference and current electrodes, using 2 mm filaments (the low W1/W2 area ratio minimised polarisation over the steel probe). An ACM instrument was used for recording electrochemical potential and current noise signals at a sampling rate of 1 reading/s.

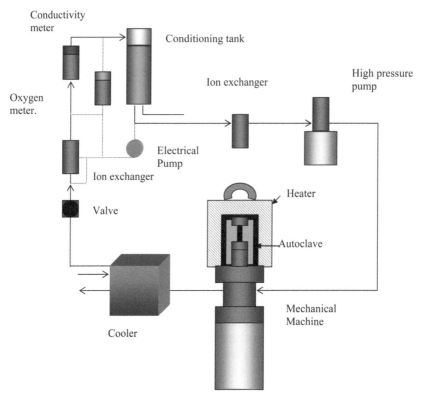

5.1 Simplified diagram of the autoclave and water conditioning loop to simulate a BWR environment

5.3   Results

Figure 5.3 shows the load–time curve and the potential and current time-series obtained for a solution annealed 304 steel. At the initial stage, before the yield point was reached, a low noise signal was obtained. This is the typical background noise in the absence of any cracking events. Figure 5.4a shows the current signal. Beyond the yield point, different shapes of current and potential transients were observed which were believed to be related to cracking events. In these time series, passivity break-down could be observed after yielding as a current spike followed by a slow recovery in both current and voltage. The first current pulse (1.7E4 s) had a short rise time, typically a few seconds long, and a slow decay; this is magnified in Fig. 5.4b. The following pulses were caused by the overlapping of four or more transients which formed a 'tooth shape' signature, such as in Fig. 5.4c. As the test continued, single high intensity transients appeared (8.3E4 s). At a later stage (12E4 s), symmetric spikes (with rise or activation time t_a similar to recovery time t_r) were obtained (Fig. 5.4d). In addition, a load transient occurred at 17E4 s due to an interruption of the electric power supply to the test system, which demonstrated the sensitivity of the technique to detect changes in the test condition, such as removing the load. The EN technique was sensitive to crack initiation and the end of the cracking stages, where the specimen failed (22E4 s).

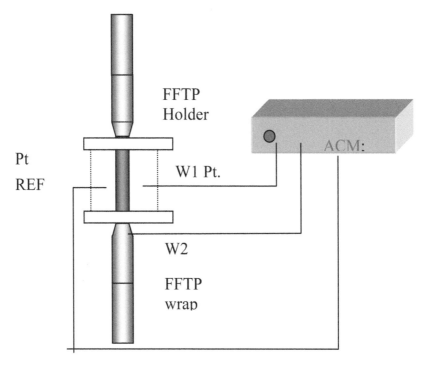

5.2 Arrangement with two platinum electrodes used to obtain EN during SS-RTs. W1 = working electrode 1 (counter 2 mm, platinum filament), W2 = working electrode 2 (tensile 304 SS probe), REF = reference electrode (2 mm of platinum filament)

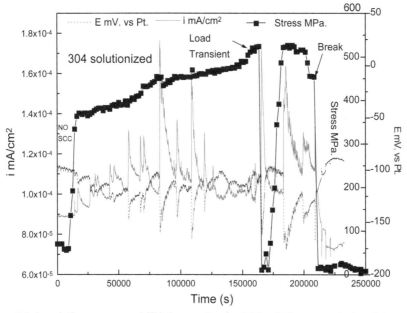

5.3 Load–time curve and EN-time series for 304 solution annealed stainless steel, at 288°C, 8 MPa, air saturated water (approximately 4000 ppb oxygen and +200 mV$_{SHE}$), outlet conductivity 0.59 μS/cm

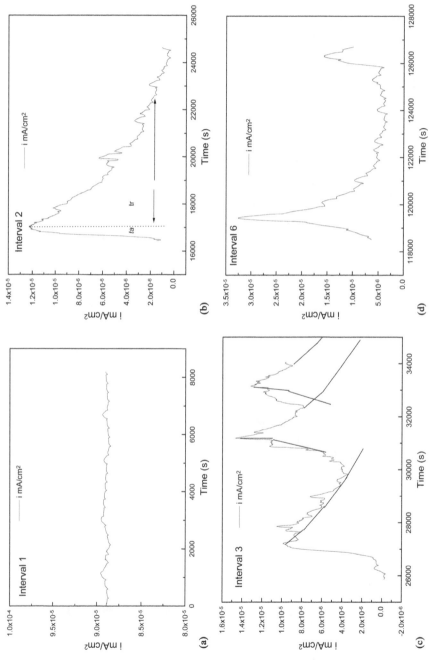

5.4 Different EN intervals of signatures in 304 solution annealed stainless steel. (a) Without transients, (b) typical SCC transient, (c) overlapping of three transients forming the 'tooth shape' and (d) symmetric spike

Figure 5.5 shows that the solution annealed material presents initial TGSCC similar to the 'fan ribs' morphology reported by Newman [9] in alpha brass or Transgranular type II with 'fan ribs line' proposed by Solomon [8] at the outer cross-section, followed by a TGSCC similar to Type I reported by Solomon [8] and then ductile fracture with microvoids in the centre of the specimen. Relating the sequence of events in the test and the fractography, the first signature of transients could be explained by the initiation of the cracking and its 'fan ribs' morphology. The single high intensity transients could be associated with the transgranular cracking, and final spikes could be associated with the ductile fracture by overload.

Figure 5.6 shows the load–time curve and time series obtained for 304 steel with a thermal sensitisation treatment. As in the previous case, a low noise signal was obtained in the period that preceded the yield point (8E3 s) in the absence of cracking. Passivity breakdown took place after the specimen reached the yield point, with a signal formed by spikes followed by recoveries characteristic of a cracking phenomenon. In this case, a quasi-periodic breakdown behaviour was observed, where breakdown recovery events are separated by periods of relatively low noise activity. During the test, breakdown events took place between crack initiation and the end of cracking, where the specimen broke (see Fig. 5.6b).

Figure 5.7 shows the fractographic analysis where the IGSCC initiation can be seen, followed by TGSCC type I [8] and finally, in the centre, ductile fracture.

Overall, the signals obtained showed more rapid activation events followed by shorter exponential recoveries than in the solution annealed case, which is in agreement with the noise signals in similar media for SCC [4,5]. Stewart et al. [5] associated the first transients with the advance of short cracks. Other transients (3E4 s) comprised a group of overlapping current pulses which finally decayed to the baseline, similar to the behaviour reported by Wells et al. [10]. These were associated with the movement of the crack across individual grain boundary facets. The single transients with high intensity (65E5 s) could be associated again with TGSCC and the symmetric spikes (105E5 s) with ductile overload fracture.

The difference in magnitude between the current density signals of the solution annealed and sensitised specimens could be attributed to small variations in the positioning of the platinum electrodes, which was done manually, or may be due to the differences in outlet conductivities, 0.59 µS/cm for solution annealed and 0.95 µS/cm for sensitised 304 SS.

The 304 solution annealed steel had a more negative potential of –98 mV versus Pt at the beginning of the test and finished at –166 mV versus Pt. The sensitised 304 steel had a potential of 41 mV versus Pt, and finished at –61 mV versus Pt, which means that this probe showed more noble behaviour (positive potential). This means that the probes were at different potentials. However, the Pt electrode is not a reference electrode, but only a 'pseudo reference' electrode and can be influenced by the environment and pH. The differences in conductivities could also have contributed to the differences of the Pt potential via an effect on the pH.

In all cases, typical rapid rise and slow recovery transients were present, but with differences in amplitude, and sometimes with overlapping of three or more transients forming a 'tooth shape'. The only different pattern was similar to the transients reported by Wells et al. [10], but this could also have been caused by overlapping of single typical SCC transients. The symmetric spikes could also be well differentiated from the SCC transients.

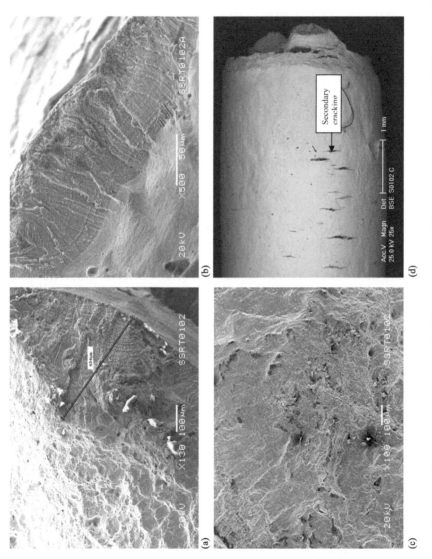

5.5 SEM fractographs of solution annealed stainless steel 304 after failure. (a) General view showing TGSCC mode of cracking and ductile fracture at the centre, (b) initiation by 'fan ribs', (c) other general view of the fracture shows 'fan ribs' + TGSCC and (d) secondary cracking and necking

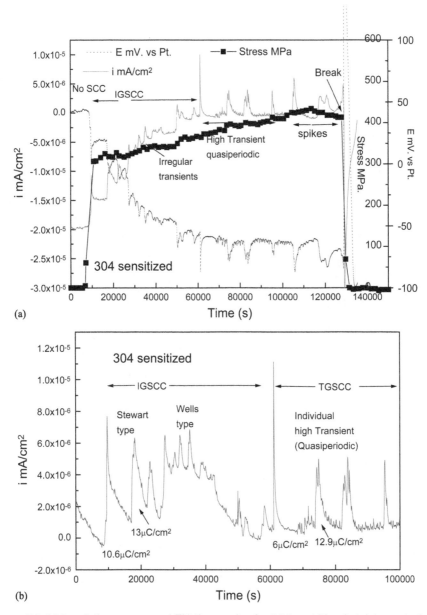

5.6 (a) Load–time curve and EN-time series for 304 sensitised stainless steel at 288°C, 8 MPa, air saturated water (approximately 4000 ppb oxygen and +200 mV$_{SHE}$), outlet conductivity 0.95 µS/cm. (b) EN-time series for 304 sensitised stainless steel at 288°C, 8 MPa, air saturated water (approximately 4000 ppb oxygen and +200 mV$_{SHE}$), conductivity 0.95 µS/cm, with trend and background removal

As mentioned by Dawson [11], anodic transients can also be the result of metal dissolution at the crack wall and therefore, similar to pit initiation. Therefore in many cases, it is difficult to discriminate between anodic SCC, pitting and crevice corrosion.

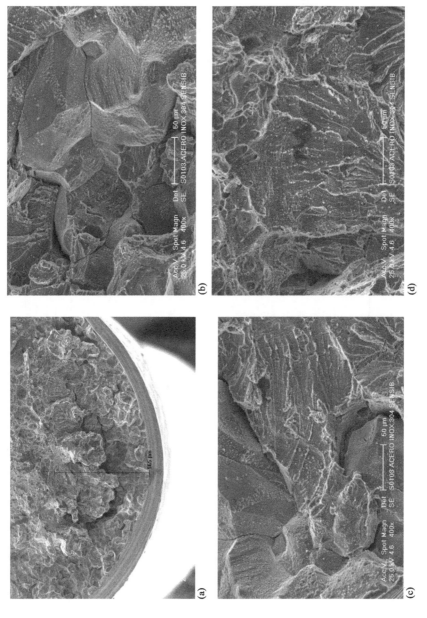

5.7 SEM fractographs of sensitised 304 stainless steel after failure. (a) General view showing IGSCC and TGSCC modes of cracking and ductile fracture at the centre, (b) initiation by IGSCC, (c) sensitised 304 showing change in fracture mode from IG SCC to TGSCC and (d) change in fracture mode from TGSCC to ductile fracture ('fan ribs' are not well defined)

An advantage of the EN technique is that it can detect events that occur sequentially, including the absence of events as shown by periods without transients; in this case, these are related to the different mechanisms of cracking, e.g. 'fan ribs' before TGSCC, or IGSCC before TGSCC, and TGSCC before ductile fracture.

Unfortunately, the amplitude of the signature also depends on differences in the experimental set-up, such as the distance between electrodes. In this case, it is difficult to associate the amplitude and the integrated 'electric dissolution charge' with the depth of the crack. For example, the larger transients with high amplitude and high dissolution charge (Table 5.3) were attributed to TGSCC in the solution annealed steel with 484 µm crack depth (see Fig. 5.5a), and large cracking with approximate 965 µm depth (see Fig. 5.7a) was attributed to IGSCC and TGSCC with small charge transients (Table 5.4).

Watanabe et al. [12] and Wells et al. [10] obtained a good relationship between the number of transients and the number of cracks and found a direct correlation between current pulses and crack initiation. Figure 5.8 shows the time series for a 304 SS tested in a SSRT at 1.5E-7 s⁻¹ and stopped at 0.44 mm extension. In this case, Fig. 5.8b and c only show one zig-zag shaped crack, which gave rise to five transients. This shows that perhaps the transients are more related to the growth and repassivation of the crack as proposed by the slip-dissolution model [2] and that one crack can produce more than one transient compared with the 'one to one' correlation proposed by Watanabe et al. [12] and Wells et al. [10].

Table 5.3 shows six time intervals selected to obtain the power density spectrum (PDS) shown in Fig. 5.9a for the solution annealed 304 SS. Figure 5.9b shows the standard deviation (STD) as a function of time for the same set of intervals. Interval one was taken from the elastic zone, without any electrochemical transients and presumably no SCC. It presents three regions, the first with a plateau zone and a

Table 5.3   Intervals selected for the time series of EN in solution annealed 304 SS

Interval	Comments	n Decay rate	T_a, s	T_r, s	Amplitude I, mA/cm²	Amplitude E, mV	Q µC
I1 0-8192 (s)	Without transients No SCC	–	–	–	–	–	–
I2 16443-24635(s)	Typically SCC Fan ribs	0.16	429	2526	1.03E-5	8	107
I3 25861-34053(s)	Overlapping tooth shape Fan ribs		–	–	1.4E-5	17.6	170
I4 81722-89914(s)	Individual transient TGSCC	0.15	389	11260	7.06E-5	48	478
I5 108539-116731(s)	Individual transient TGSCC	0.21	338	7635	5.43E-5	48	507
I6 118504-126696(s)	Spikes (ductile)	0.38	768	1537	2.7E-5	30	78

Table 5.4   Intervals selected for the time series of EN in sensitised 304 SS

Interval	Comments	n Decay rate	T_a, s	T_r, s	Amplitude i mA/cm²	Amplitude E mV	Q µC
I1 0-8192 (s)	Without transients No SCC	–	–	–	–	–	–
I2 8619-16731(s)	Typically SCC IGSCC	0.3	182	949	6.19E-6	26	28.2
I3 16811- 25003(s)	Typically SCC IGSCC	0.17	1076	2472	4.6E-6	23	29.64
I4 27003- 35195(s)	Typically SCC IGSCC	–	508	1654	4.3E-6	20	
I5 37990- 46182(s)	Overlapping (Well's type) IGSCC Movement of the crack across individual grain boundary facets	0.14	–	–	1E-6	6.8	
I6 61007 69199(s)	Individual transient TGSCC	0.51	93	3341	5.43E-5	48	16.92
I7 74080 82272(s)	Individual transient TGSCC	0.31	1143	4484	3.6E-6	31	12.9
I8 104512 112704(s)	Spikes (ductile)	0.49	631	10388	6.2E-6	16.92	20.02

slope of –0.375 log(PDS i)/decade, the second a roll-off slope of –2.20 log(PDS i)/decade and a third region with a slope of –3.08 log(PDS i)/decade. In the interval with transients associated with 'fan ribs', corrosion processes should be indicated by the increasingly negative or sharper roll-off slopes and higher noise amplitudes. The maximum amplitude values reached in interval four were associated with TGSCC. After these, the amplitudes decreased, indicating that the SCC process was less severe. This is in agreement with the increase in the STD versus time from interval one to four and the decrease in the STD in intervals five and six. The relatively small transients correlated with the initial 'fan ribs' mode of fracture, the increase in the signals with the TGSCC mode and the decrease in the signals with ductile fracture and necking.

In Table 5.3, the values of charges associated with the noise transients are in agreement with those obtained by Stewart et al. [5] from 10 to 200 µC in sensitised 304 SS under similar conditions.

Table 5.4 shows a set of intervals selected to obtain the PDS plot shown in Fig. 5.10a for the sensitised 304 SS. Figure 5.10b shows the STD as a function of time

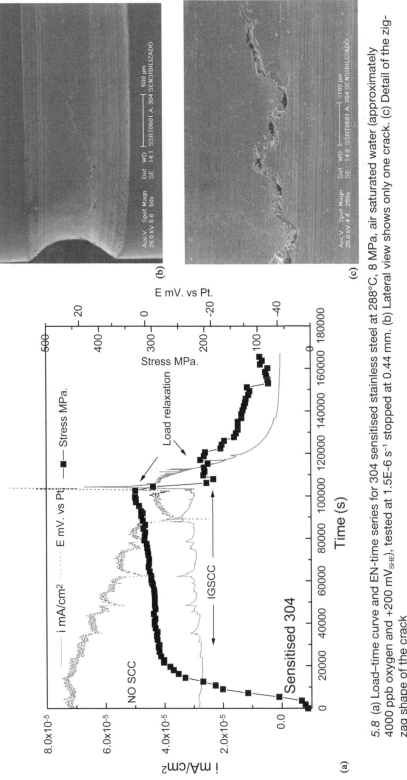

5.8 (a) Load–time curve and EN-time series for 304 sensitised stainless steel at 288°C, 8 MPa, air saturated water (approximately 4000 ppb oxygen and +200 mV$_{SHE}$), tested at 1.5E-6 s^{-1} stopped at 0.44 mm. (b) Lateral view shows only one crack. (c) Detail of the zig-zag shape of the crack

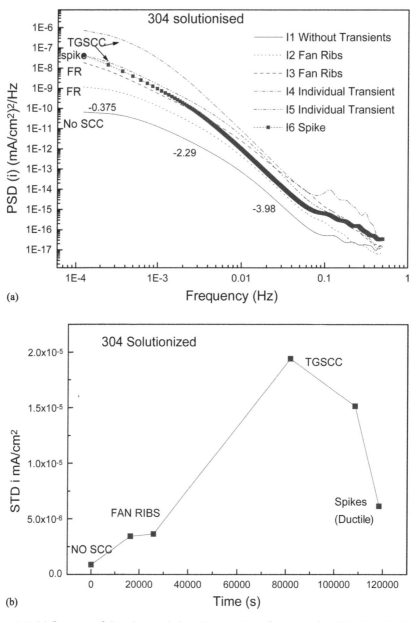

5.9 (a) Spectra of time intervals in a time series of current densities in solution annealed 304 stainless steel. (b) STD vs. time for the same time intervals in solution annealed 304

for the chosen set of intervals. In Fig. 5.10a, interval one, which is related to the time before SCC initiated, also shows three regions. The first one with a slope of –0.25 log(PDS i)/decade, the second one of –2.21 log(PDS i)/decade and finally a region of –4.51 log(PDS i)/decade. In the intervals associated with SCC, the process should be indicated by the increasingly negative or sharper roll-off slopes and higher noise amplitudes. In this case, there is no direct relationship between STD and PDS

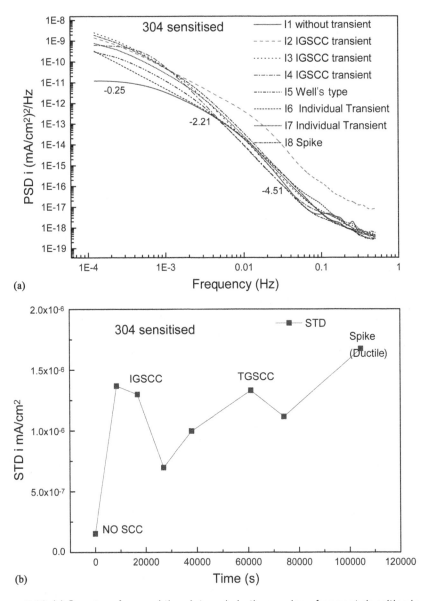

5.10 (a) Spectra of several time intervals in time series of current densities in sensitised 304 stainless steel. (b) STD vs. time for the same time intervals in sensitised 304

spectrum. This is because IGSCC was beginning with small amplitude transients which increased, but after that the intensity decreased in irregular transients associated with the arrest of this fracture mode. Later, the current density signal increased again associated with the TGSCC transient, and then finally the failure mode was more abrupt and ductile fracture produced big spikes.

In Table 5.4, the values of charges associated with transients are in agreement with those obtained by Wells et al. [10] from 1 to 25 μC in sensitised 304 SS, in dilute thiosulphate solutions at room temperature.

The intensity of the pulses, and the dissolution charge Q, depend on the differences in potential between the steel and the Pt pseudo reference electrode. For example, the solution annealed 304 SS had a larger potential difference and produced larger pulses with more charge Q. The sensitised 304 SS had smaller differences in potential and produced low amplitude pulses with a small charge Q. Decay behaviour of current or current density in a repassivation process is usually expressed by the exponential form:

$$i = i_o \left(\frac{t}{t_o} \right)^{-n}$$ [3.1]

where t is the time for new exposure of a bare metal surface, i_0 the peak current density and t_0 is the lead time for the current to start decaying. The values of n for decay kinetics obtained by log–log plots of the decay sections of the transients, in the region from 100 to 1000 s, show that for 304 solution annealed steel in intervals 2 and 4 (associated with 'fan ribs' and TGSCC), $n = 0.16$ and 0.15. For interval 5 (TGSCC), $n = 0.21$ and for interval 6 (spike due to ductile fracture), $n = 0.38$. By comparison, for sensitised 304 steel, in interval 2, $n = 0.3$, for interval 3, $n = 0.17$ and for the decay after irregular transients in interval 5, $n = 0.14$ (all attributed to IGSCC). For interval 6 associated with TGSCC, $n = 0.51$, but interval 7 has $n = 0.33$. For a spike in interval 8, $n = 0.49$. This shows that the repassivation rate is similar for 'fan ribs' and IGSCC, $n = 0.15$, and between some IGSCC and TGSCC transients, $n = 0.33$. Some of the transients associated with TGSCC had slower repassivation rates with $n = 0.41$. Spikes attributed to ductile fracture had the highest n value between 0.38 and 0.49.

The similar values of n between different types of cracking in this case show that it is difficult to distinguish the two cases (IGSCC and TGSCC) without supporting information from other techniques. Furthermore, it is possible that overlap of signals occurs between the main crack and secondary crack, or that the main crack is in the TGSCC mode and a secondary crack with 'fan ribs' mode occurs simultaneously and the total signal is then a mixture of the two cracking modes.

The types of fracture observed were clearly related to the sensitisation. If the material was heavily sensitised, the corrosion rate of the grain boundaries was higher and the fracture mode became intergranular. For the material with only slight sensitisation, a crack propagated along some crystallographic planes where slip and favoured by dynamic strain. When the corrosion rate of grain boundaries was comparable to that of slip planes within a grain matrix, the fracture mode became a mixture of trans- and intergranular. If the environment was too mild for the slip surface to corrode preferentially, a crack could not propagate along the slip surface and the fracture mode became ductile.

The current transients may be more related to the repassivation of a bare surface than to the way this bare surface is created, because in all of the cases, the increase in current is linear (only in a few cases are symmetric spikes increasing exponentially). However, the repassivation kinetics might be different along grain boundaries and in the grain matrix and therefore depend on the cracking mechanism. The corrosion rate and therefore the repassivation kinetics occur slowly at grain boundaries due to the Cr depleted zone in IGSCC or by hardening due to enrichment of impurity elements, such as phosphorus, sulphur and silicon, and the decay rate is small. Watanabe et al. [4] found $n = 0.3$ for IGSCC in similar media. For TGSCC, the kinetics are related more to the strain rate and the slip of planes. There is no big chemical difference between these planes and the grain matrix and there is also sufficient Cr to form a passive layer. The decay rate is expected to be higher under these conditions.

The 'fan ribs' mechanism is more complicated to explain, but there is a front at 180° that advances in a radial and then in a circumferential way. This requires that the crack stops growing in depth, but continues to advance laterally and therefore the repassivation occurs more slowly and the decay rate is low.

There is not a definitive method to distinguish between the different cracking modes by EN signals because both produce bare surfaces, but the repassivation processes are different in IGSCC and TGSCC.

The differences in potential transients are more evident than the current transients as shown in Tables 5.3 and 5.4. It is possible to associate the potential amplitude with the type of transitory (e.g. 8–18 mV to 'fan ribs', 31–48 mV to TGSCC, and 20–26 mV to IGSCC, but there is no rule) and symmetric spikes (attributed to ductile fracture) are better distinguished by the signal shape.

5.4   Summary and conclusions

EN measurements were performed during slow strain rate tests with 304 stainless steel in oxidising high-temperature water at 288°C.

Characteristic EN signals were obtained corresponding to crack initiation and propagation under similar conditions to those reported favouring SCC. A number of current transients were observed from the austenitic stainless steel specimens under constant straining in the simulated BWR environment. No transients were detected before the yield point. The shape and polarity of the transients indicated that they corresponded to slip-dissolution events.

For the solution annealed 304 steel, multiple cracking occurred with a 'fan ribs' morphology, followed by TGSCC and finally ductile fracture in the centre section of the specimen.

The cracking mode for the sensitised 304 condition was intergranular, followed by a transition to transgranular cracking and finally ductile failure in the centre section of the specimen.

It turned out that one of the great advantages of the EN technique, compared to other techniques, is that it can detect fracture events that occur sequentially, separated by low-noise intervals. The transients can be related to the different cracking modes, e.g. 'fan ribs', TGSCC, IGSCC and ductile fracture.

Acknowledgements

The authors acknowledge the contribution of Juan Andrés Aguilar Torres for the handling of the autoclave and loop at the High Temperature Corrosion Laboratory, CONACYT and the IAEA funding is also acknowledged.

References

1. R. M. Horn, G. M. Gordon, F. P. Ford and R. L. Cowan, *Nucl. Eng. Des.*, 174 (1997), 313–325.
2. F. P. Ford, *Corrosion*, 52(5) (1995), 375.
3. D. D. Macdonald, C. Liu and M. Manahan, 'Electrochemical noise measurements of carbon and stainless steels in high subcritical and supercritical aqueous environments',

in *Electrochemical Noise Measurement for Corrosion Applications*, STP 1277, ed. J. F. Kearns, J. R. Scully, P. R. Roberge and D. L. Reichert. ASTM, Philadelphia, PA, 1996.

4. Y. Watanabe, T. Shoji and T. Kondo, 'Electrochemical noise characteristics of IGSCC in stainless steels in pressurized high-temperature water', in *Corrosion 98*, Paper No. 129. NACE, San Diego, CA, 1998.

5. J. Stewart, D. B. Wells, P. M. Scott and D. E. Williams, *Corros. Sci.*, 33(1) (1992), 73–88.

6. P. L. Andresen, P. W. Emigh, J. Hickling, D. F. Taylor, J. M. Burguer, R. M. Horn and R. Pathania, 'Detection of stress corrosion crack initiation in the BWR environment by electrochemical noise', in *8th International Symposium on Environmental Degradation of Materials in Nuclear Power System-Water Reactors*, 622–631. NACE, AMS, TMS, 1997.

7. L. H. Wang, J. J. Kai, C. H. Tsai and C. Fong, *J. Nucl. Mater.*, 258–263 (1998), 2046–2053.

8. H. D. Solomon, *Corrosion*, 40(9) (1984), 493–506.

9. R. Newman, 'Stress-corrosion cracking mechanisms', in *Corrosion Mechanisms in Theory and Practice*, Chapter 10, ed. P. Marcus and J. Oudar. Marcel Dekker Inc., USA, 1995.

10. D. B. Wells, J. Stewart, R. Davidson, P. M. Scott and D. E. Williams, *Corros. Sci.*, 33(1) (1992), 39–71.

11. J. L. Dawson, 'Electrochemical noise measurement. The definitive in-situ technique for corrosion applications?', in *Electrochemical Noise Measurement for Corrosion Applications*, STP 1277, ed. J. R. Kearns, J. R. Scully, P. R. Roberge, D. L. Reichert and J. L. Dawson. ASTM, Philadelphia, PA, 1996, 3–35.

12. Y. Watanabe, T. Shoji and T. Kondo, 'Current and potential fluctuation characteristics in IGSCC processes of stainless steels', in *Corrosion 98*, Paper No. 376. NACE, San Diego, CA, 1998.

6

Stochastic approach to electrochemical noise analysis of SCC of Alloy 600 SG tube in caustic environments at high temperature*

Sung-Woo Kim, Hong-Pyo Kim, Seong-Sik Hwang,
Dong-Jin Kim, Joung-Soo Kim, Yun-Soo Lim,
Sung-Soo Kim and Man-Kyo Jung

Korea Atomic Energy Research Institute, 1045 Daedeok-Daero, Yuseong-Gu, Daejeon 305-353, Republic of Korea

kimsw@kaeri.re.kr

6.1 Introduction

Although Alloy 600 has been used as the steam generator (SG) tubing material in pressurised water reactors (PWRs) due to its high corrosion resistance, many types of corrosion have occurred in environments containing some oxidising impurities in SG sludge piles. In particular, Pb is a well-known impurity that assists in stress corrosion cracking (SCC) of SG tubing in caustic environments [1–7]. Many authors have reported on the cracking modes of SCC in Pb-contaminated solutions and the role of Pb on the passive films formed on Alloy 600 to explain the mechanism, and the crack growth kinetics of PbSCC in various aqueous environments [2–7]. However, there are few reports that successfully distinguish between the initiation and propagation stages of PbSCC because it is difficult to detect them separately with conventional techniques.

Since electrochemical noise (EN), a fluctuation of the electrochemical potential and current, has been observed experimentally to be associated with localised corrosion processes, it has become a useful tool for monitoring localised corrosion such as pitting corrosion, crevice corrosion and SCC [8–16]. It has been reported that this technique is powerful enough to distinguish between different types of localised corrosion using a stochastic theory with the probabilities of the EN parameters such as the frequency of events, noise resistance, average charge of events [10–12] and mean free time-to-failure [14–16].

The purpose of the present work is to examine the initiation and propagation processes of PbSCC in Alloy 600 using the EN technique and stochastic theory. The electrochemical potential noise (EPN) and electrochemical current noise (ECN) were measured simultaneously from Alloy 600 C-ring specimens stressed to 150% of their room temperature yield strength (YS) in caustic solutions containing oxidising impurities. The EN parameters such as the frequency of events and the mean free time-to-failure were obtained from the EN data, and then the shape parameters were calculated from the Weibull probability function. Analyses were performed for test solutions with different impurities. The experimental data and calculations are discussed in terms of the stochastic characteristics of the uniform corrosion, initiation and propagation of PbSCC.

6.2   Experimental

Alloy 600 SG tubing (Valinox Heat No. NX8527) was used for this study: 22.23 mm outer diameter, 1.27 mm thickness. The chemical composition of the materials is given in Table 6.1. The tubing was pilgered by the manufacturer, and heat treated in our laboratory at 920°C for 15 min in an Ar-filled quartz capsule in a furnace, followed by water quenching, to simulate low-temperature mill-annealing (LTMA). The C-ring specimens were fabricated from the Alloy 600 SG tubing in accordance with the ASTM G38 standard. The outer diameter (OD) surface of the specimens was ground with #1000 emery paper, and then cleaned with ethanol and water in sequence. The C-ring specimen was stressed to 150% of its room temperature YS at the apex using an Alloy 600 bolt and nut.

The test environments were an aqueous solution of 40 wt.% NaOH solution containing 0.01 wt.% PbO (leaded caustic solution), plus up to 500 ppm CuO. It has been reported that highly caustic conditions can be developed in the heated crevices of PWR SGs where Alloy 600 may become susceptible to SCC [17]. The Pb and Cu species are frequently found in SG sludge piles [17]. In the present investigation, CuO was added to the leaded caustic solution to accelerate PbSCC. CuO is a well-known oxidising species that increases the corrosion potential and promotes PbSCC [17,18]. It has been reported that an acceleration of PbSCC can also be accomplished by the potentiostatic method by maintaining the electrochemical potential in the active-passive range, 0.1–0.2 V anodic to the open circuit potential (OCP) for Alloy 600 in caustic environments [18]. The potentiostatic method is not an option for the present EPN experiments.

Figure 6.1 is a schematic of the C-ring test facility and the EN experiments. The EPN and the ECN measurements were carried out with a Zahner IM6e equipped with a Zahner NProbe. The experiments were performed on a galvanically-coupled pair of stressed (150% YS) and unstressed C-ring specimens. The galvanic coupling was accomplished using the zero-resistance ammeter (ZRA) mode of the IM6e. The reference electrode was an external Ag/AgCl/KCl (0.1 M) electrode, specially designed and constructed for high-temperature and pressure applications (Toshin-Kogyo Co., UH type). For reliable measurements of the electrochemical potential, the reference electrode was reassembled and recalibrated frequently during the interruption of the test series described below. The current transients between the two specimens of the pair, and the potential transients of the stressed specimen versus the reference electrode were monitored simultaneously. The sample interval and the frequency bandwidth were 0.5 s and 0.001–1 Hz, respectively. The power spectral density and the frequency of events were calculated from each set of time records that consisted of 2048 data points acquired for 1000 s.

Table 6.1   Chemical composition of the Alloy 600 materials (in wt.%)

C	Mn	Si	P	S	Ni	Cr	Ti	Al	Co	Cu	Fe
0.024	0.23	0.13	0.005	<0.001	74.90	15.37	0.28	0.20	0.016	0.009	8.84

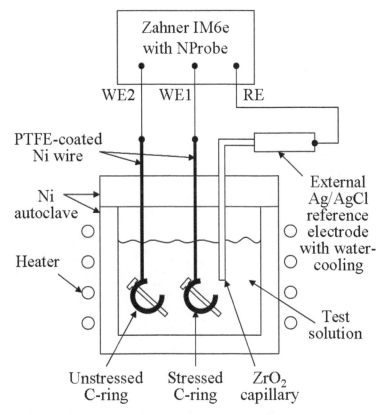

6.1 Schematic of the C-ring test system and the EN experiment (from S. W. Kim and H. P. Kim, *Corr. Sci.,* 51(1) (2009), 191, with permission)

The tests were performed at 290°C for a total accumulated immersion time of 400 h, and consisted of three phases of immersion in sequence; (1) 264 h in the leaded solution (40 wt.% NaOH with 0.01 wt.% Pb), (2) 14 h in the leaded solution plus 200 ppm CuO, and (3) 122 h in the leaded solution plus 500 ppm CuO. The 200 ppm of CuO was added at 264 h, and then an additional 300 ppm at 278 h. The addition of CuO caused an increase in the electrochemical potential by 0.15–0.2 V. The tests were interrupted a number of times and the surfaces of the specimens were examined for cracking using an optical stereo-microscope, at accumulated immersion times of 50, 144, 264, 278 and 400 h. The test solution was de-aerated with 99.99% nitrogen gas for 20 h before the tests began.

After the entire immersion test, the specimens were chemically etched with a solution of 2% bromine + 98% methanol, and then examined with a scanning electron microscope (SEM, JEOL JSM-6360) equipped with an energy dispersive spectrometer (EDS, Oxford-7582).

6.3   Results and discussion

6.3.1   Microscopic analysis

Figure 6.2(a) is the optical stereo-micrograph of the OD surface of the stressed C-ring apex immersed for 264 h in the leaded caustic solution and Fig. 6.2(b) is an oblique

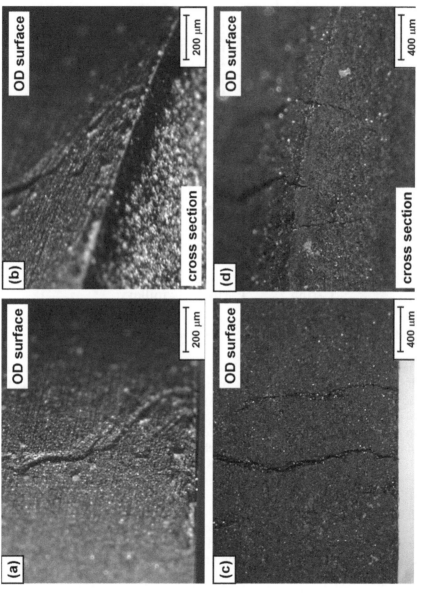

6.2 Optical stereo-micrograph of (a) the OD surface and (b) the corresponding oblique view of the C-ring apex immersed in the leaded caustic solution at 290°C for 264 h, and (c) the OD surface and (d) the corresponding oblique view of the C-ring apex after the entire immersion test at 290°C for 400 h (from S. W. Kim and H. P. Kim, *Corr. Sci.*, 51(1) (2009), 191, with permission)

view corresponding to Fig. 6.2(a). Figs. 6.2(a) and (b) show a feature that may indicate a surface film breakage or local corrosion. However, extensive examination of the cross-section of the specimen ruled out any cracking. After the entire immersion test in the leaded caustic solution with CuO for 400 h, several cracks were propagated as shown in Figs. 6.2(c) and (d).

Figure 6.3 presents the SEM micrograph of the cross-section of the C-ring specimen after the entire immersion test. The crack was propagated in an intergranular (IG) mode and the crack mouth and tip were covered with a surface film. The chemical composition of the surface film was analysed by energy dispersive spectroscopy (EDS), and summarised in Table 6.2, along with the chemical composition of the Alloy 600 matrix. Pb was detected in the oxide film at both the crack mouth and the crack tip. This is typical of PbSCC in Alloy 600 in leaded caustic solutions [3–7].

After the entire immersion test, many cracks were initiated at micron-scale through the grain boundary at the C-ring apex surface as shown in Fig. 6.4. Those crack initiations were not able to be detected at the intermittent crack inspection using optical stereo-microscopy.

6.3.2   Electrochemical noise analysis

Figure 6.5 demonstrates the time records of the EPN and ECN measured from the C-ring specimens at 290°C for 400 h with different additives. There were abrupt changes in the EPN or ECN at an interruption for crack inspection (points A, B, C, D and E in Fig. 6.5). These may be attributed to fluctuations of the test environment such as temperature, pressure and water chemistry at the interface between the solution and specimen exposed to the room temperature atmosphere for crack inspection.

There were also rapid increases in the EPN by 150 and 50 mV in sequence due to the increase in the corrosion potential of the specimens with additions of 200 and 300 ppm CuO, respectively, to the leaded caustic solution. It is well known that the addition of CuO to a caustic solution as the sludge constituent in the SG, increases the corrosion potential of Alloy 600 and hence its SCC susceptibility at high temperature [17,18].

Separately, there were typical potential decreases in the EPN concurrent with the current rises in the ECN, which have generally been observed during localised corrosion such as pitting corrosion, crevice corrosion, intergranular corrosion and SCC [8–13]. Figures 6.6(a) and (b) present in detail these typical EPN and ECN values recorded for the leaded solution for the time period from 720 000 to 770 000 s, and for that solution plus CuO for the time period from 1 330 000 to 1 436 000 s, respectively. The current transient accompanied by a random potential drop in Fig. 6.6(a) revealed a repetitive increase and decrease in a stepwise manner at a lower amplitude and at a shorter time interval compared to the current transient exhibiting a rapid rise and a slow fall in Fig. 6.6(b).

From the microscopic analysis, it is easily anticipated that the current transient behaviour in Fig. 6.6(a) is mainly due to the local breakdown and repassivation of the surface oxide film; the increase in the current corresponding to the potential decrease means a local breakdown of the oxide film and hence an exposure of the metal surface, whereas the decrease in the current accompanied by the potential recovery indicates a repair of the oxide film. Similar transient behaviours have been

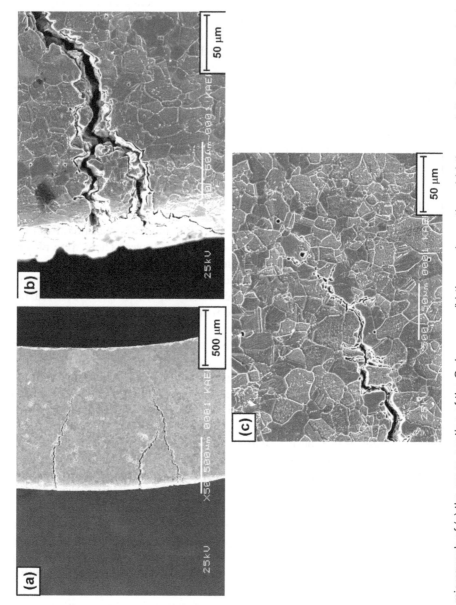

6.3 SEM micrograph of (a) the cross-section of the C-ring apex, (b) the crack mouth and (c) the crack tip after the entire immersion test at 290°C for 400 h (from S. W. Kim and H. P. Kim, *Corr. Sci.*, 51(1) (2009), 191, with permission)

Table 6.2 Chemical composition of the matrix and surface film of the crack mouth and tip after a whole immersion test for 400 h at 290°C (in wt.%)

Position	O	Ti	Cr	Mn	Fe	Ni	Pb
Matrix	ND	0.22	14.49	0.47	7.96	76.05	ND
Crack mouth	8.39	0.19	13.80	0.17	6.97	70.40	0.94
Crack tip	8.25	0.25	13.95	0.25	7.26	70.04	0.40

ND, not detectable.

reported for a passivity breakdown triggering the initiation of localised corrosion [9,13,19].

On the other hand, the current transient in Fig. 6.6(b) revealed a more localised behaviour, that is, a larger amplitude with a longer time interval, indicating a larger exposure of the metal surface with a lower event frequency, compared to that of the current transient in Fig. 6.6(a). Anita et al. [13] reported a similar difference in the potential and current transients for the propagation of SCC from those for metastable pitting. Consequently, it is strongly suggested that the potential decrease in the EPN concurrent with the current rise in the ECN in Fig. 6.6(b) is due to the propagation of PbSCC, while that in Fig. 6.6(a) is attributable to the initiation of PbSCC.

Figures 6.7(a) and (b) present plots of the power spectral density (PSD) vs. the frequency calculated from each time record by a fast Fourier transform (FFT) algorithm for the time period from 720 000 to 770 000 s and from 1 330 000 to 1 436 000 s, respectively. In Fig. 6.7(a), it is clearly seen that the PSD obtained at points A, C and D, where the EPN reveals the potential decrease accompanied by the current increase in the ECN, is higher than that obtained at points B and E, where the EPN reveals the potential recovery corresponding to the current decrease in the ECN. Similar behaviours were found in Fig. 6.7(b), that is, the PSD obtained at points A, B, D and E is higher than that obtained at point C. The PSD increases remarkably at a low frequency limit in Figs. 6.7(a) and (b), which strongly indicates the increase in the number of localised corrosion events, that is, the initiation and propagation of PbSCC, respectively, in the present work.

It has been proposed that a shallow roll-off slope of the power spectra density plot of –20 dB/decade or less is indicative of pitting corrosion whereas a steep roll-off slope of more than –20 dB/decade represents general corrosion or passivation [20,21]. In contrast to the previous results, the roll-off slope for the propagation of PbSCC is steeper than the slope for the uniform corrosion in this work: the roll-off slopes of the power density spectrum were –18 dB/decade for uniform corrosion (B and E in Fig. 6.7(a) and C in Fig. 6.7(b)), –20 dB/decade for initiation of PbSCC (A, C and D in Fig. 6.7(a)) and –22 dB/decade for propagation of PbSCC (A, B, D and E in Fig. 6.7(b)). Similarly, Fukuda and Mizuno [22] and Anita et al. [13] reported a steeper roll-off slope for pitting corrosion of stainless steels than the slope for passivation. Since the type of corrosion cannot be reliably distinguished on the basis of the roll-off slope [10,13], it will not be considered as indicative of localised corrosion in this work.

Based on a shot-noise theory [10–12], the frequency of events f_n of the localised corrosion is estimated for each time record as given by

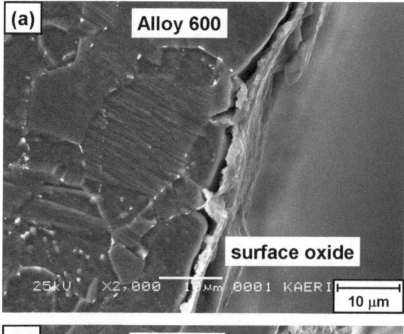

6.4 SEM micrograph of the cracks initiated through the grain boundary at the C-ring apex surface in cross-sectional view after the entire immersion test at 290°C for 400 h

$$f_n = \frac{B^2}{\Psi_E A} \qquad [6.1]$$

where B is the Stern–Geary coefficient, Ψ_E the PSD value of the EPN obtained by averaging several low-frequency points using the FFT algorithm and A represents

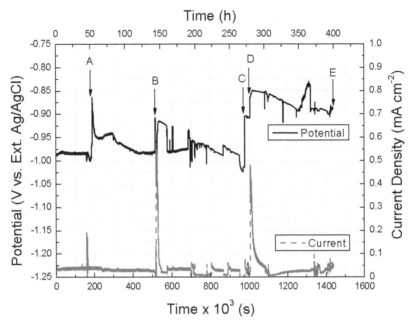

6.5 Time records of EPN and ECN measured from the stressed C-ring specimens at 290°C for 400 h with different additives. 200 ppm of CuO was added at 264 h (C), and then an additional 300 ppm at 278 h (D). The tests were interrupted a number of times and the surface of the specimen was examined for cracking at points A, B, C, D and E (from S. W. Kim and H. P. Kim, *Corr. Sci.*, 51(1) (2009), 191, with permission)

the exposed electrode area. From a set of f_n calculated from the PSD plots according to equation 6.1, the cumulative probability $F(f_n)$ at each f_n is determined numerically by a mean rank approximation [10]. In this work, to understand the stochastic characteristics of PbSCC, the probability of f_n was analysed using the Weibull distribution function [14–16,23]. The Weibull distribution function, one of the most commonly used cumulative probability functions for predicting a life and failure rate, is expressed as

$$\ln\{\ln[1/(1-F(t))]\} = m\ln t - \ln n \qquad [6.2]$$

where t is the mean free time which corresponds to $1/f_n$, and m and n represent the shape and scale parameters, respectively.

Figure 6.8 depicts the plots of $\ln\{\ln[1/(1-F(t))]\}$ vs. $\ln t$ determined numerically from the sets of f_n which were calculated for the same time period as Figs. 6.6(a) and 6(b). For a comparison, the plot of $\ln\{\ln[1/(1-F(t))]\}$ vs. $\ln t$ obtained from the EN measurement of two identical unstressed Alloy 600 wire specimens in a 40 wt.% NaOH solution without impurities at 290°C is also given in the figure. It was observed that the unstressed specimen in the 40% NaOH solution revealed uniform corrosion without any cracking. In Fig. 6.8, the shift of the mean free time t to a higher value, that is, the shift of f_n to a lower frequency, was found in the plots obtained from the stressed specimens, compared to that obtained from the unstressed specimen where only uniform corrosion occurred. This also means that there were localised corrosion events occurring in the stressed specimens.

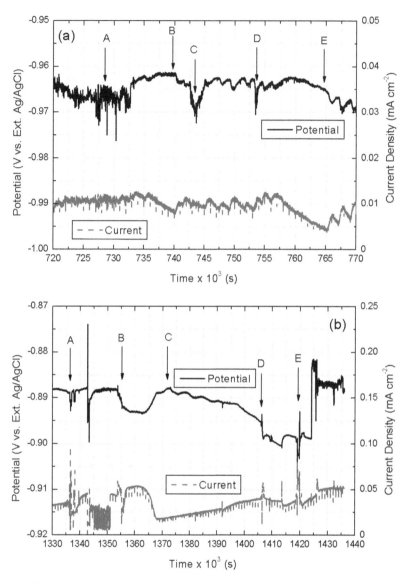

6.6 Time records of EPN and ECN measured from the stressed C-ring specimens (a) in the leaded caustic solution from 720 000 to 770 000 s, and (b) in that solution plus CuO from 1 330 000 to 1 436 000 s, respectively (from S. W. Kim and H. P. Kim, *Corr. Sci.*, 51(1) (2009), 191, with permission)

In previous works [14–16,23–25], the conditional event generation rate $r(t)$ of a localised corrosion was introduced as a failure rate using the values of m and n as

$$r(t) = \frac{m}{n} t^{m-1} \qquad [6.3]$$

Depending on the corrosion mechanism, the value of $r(t)$ was regarded as the rate of localised corrosion events such as pit generation [24], pit growth [25] and propagation of SCC [16]. In a similar way, to employ the aforementioned concept in this work,

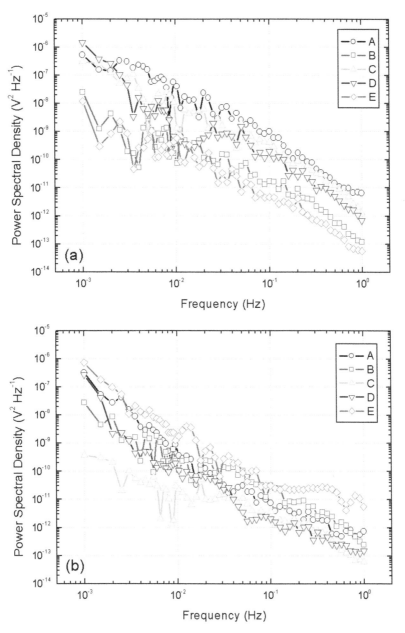

6.7 Plots of the power spectral density (PSD) vs. the frequency calculated from each time record of the EPN measured from the stressed C-ring specimens (a) in the leaded caustic solution from 720 000 to 770 000 s, and (b) in that solution plus CuO from 1 330 000 to 1 436 000 s, respectively (from S. W. Kim and H. P. Kim, *Corr. Sci.*, 51(1) (2009), 191, with permission)

the value of m was determined from Fig. 6.8 by a linear curve fitting method and is given in Table 6.3. For the stressed specimens, the value was calculated from the plots where the probability deviates considerably from that of the unstressed specimen. Bearing in mind that the unstressed specimen revealed uniform corrosion without

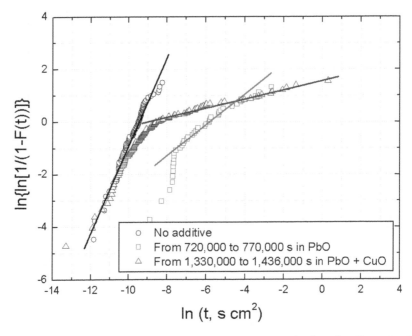

6.8 Plots of ln{ln[1/(1 − F(t))]} vs. ln*t* determined numerically using the sets of f_n for the stressed C-ring specimens in the same time period as Figs. 6.6(a) and (b). The plot obtained from the unstressed wire specimen in a 40 wt.% NaOH solution at 290°C is also given in the figure (from S. W. Kim and H. P. Kim, *Corr. Sci.*, 51(1) (2009), 191, with permission)

Table 6.3   The value of the shape parameter *m* determined in various solutions

Solution	*m*
40 wt.% NaOH	1.46 ± 0.02
40 wt.% NaOH + 0.01 wt.% PbO	0.59 ± 0.02
40 wt.% NaOH + 0.01 wt.% PbO + 500 ppm CuO	0.17 ± 0.01

any cracking, the *m* value of 1.46 can be regarded as a kinetic parameter mainly related to uniform corrosion. From the microscopic and EN analysis, the *m* value of 0.59 for the stressed specimen in the leaded caustic solution can be regarded as a kinetic parameter mainly related to the initiation of PbSCC, and the *m* value of 0.17 for the stressed specimen in the leaded caustic solution plus CuO can be assumed to be the kinetic parameter closely related to the propagation of PbSCC.

There are few reports in the literature on the stochastic approach for the localised corrosion of Alloy 600 materials at high temperature. For stainless steels, Shibata [19,26] reviewed the stochastic studies on a passive film breakdown triggering the initiation of localised corrosion. He analysed the stochastic characteristics of the local breakdown and repair of the passive film using the Poisson stochastic theory by assuming a completely random occurrence. For the Poisson process, the value of *m* in the Weibull probability is unity. In a real situation, however, there is a distortion in the Poisson probability, and hence, a deviation of the value of m from unity, caused by a difference in many experimental conditions such as the chemistry of the test solution, the applied potential and the interactions between localised events.

On the other hand, Pyun et al. reported that the value of m increases from 0.7 up to unity with a decreasing applied potential [23] for pit generation at the surface of Al alloy at room temperature, and the value of m decreases from 0.4 to 0.1 with the addition of SO_4^{2-} and MoO_4^{2-} ions to a 0.1 M NaCl solution [14] and with the addition of SO_4^{2-}, NO_3^- and PO_4^{3-} ions to a 1 M HCl solution [15] for pit initiation at the surface of pure Al at room temperature. Interestingly, they also reported that the value of m is 0.33 for the SCC propagation of Alloy 600 in a neutral solution containing PbO at room temperature [16].

Figure 6.9 depicts the plots of $\ln\{\ln[1/(1-F(t))]\}$ vs. $\ln t$ determined numerically from the sets of f_n for all of the time records measured from the stressed specimens in the leaded caustic solution, and in that solution plus CuO. The plot obtained for the leaded caustic solution consists of three parts with different m values of 1.70, 0.78 and 0.16 for a short, medium and long mean free time period, respectively, whereas the plot obtained for the leaded caustic solution plus CuO is divided into two main parts with different slopes, that is, m values of 1.51 and 0.18 for a short and long mean free time period, respectively.

From the results that the m values of the stressed specimens for a short mean free time period are similar to the m value of the unstressed specimens where only uniform

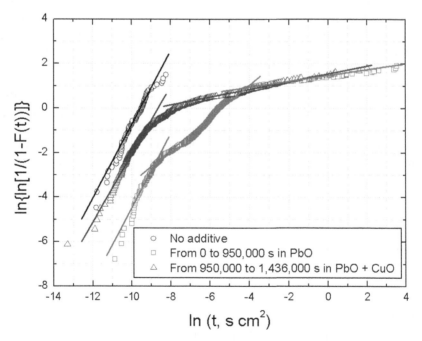

6.9 Plots of $\ln\{\ln[1/(1-F(t))]\}$ vs. $\ln t$ determined numerically using the sets of f_n for the stressed C-ring specimens for the whole time records measured from the stressed specimens in the leaded caustic solution and in that solution plus CuO. The plot obtained from the unstressed wire specimen in a 40 wt.% NaOH solution at 290°C is also given in the figure (from S. W. Kim and H. P. Kim, *Corr. Sci.*, 51(1) (2009), 191, with permission)

corrosion occurred, it is confirmed that uniform corrosion also occurs in the stressed specimens exposed to leaded caustic solution, and to that solution plus CuO.

For the stressed specimen in the leaded solution, there is another localised corrosion event with m value of 0.78 for the medium mean free time period, closely related to the local breakdown and repassivation of the surface oxide film, that is, the initiation of PbSCC. For the stressed specimen in the leaded solution plus CuO, the localised corrosion events with the m value of 0.18 would correspond to the propagation of PbSCC. Interestingly, the plot of the stressed specimen in the leaded caustic solution also revealed the long mean free time period with m value of 0.16. This indicates that there may also be a micro-scaled growth of the cracks at the surface of the specimen which is hardly detectable during crack inspection in this work. From the results, it is concluded that the EN analysis technique is applicable to distinguish between uniform corrosion, and initiation and propagation of PbSCC at high temperature.

6.4   Concluding remarks

From the microscopic and EN analysis of Alloy 600 SG tube materials in the simulated caustic solution environment of SG sludge piles at high temperature, it is strongly indicated that the random potential decrease with shorter time interval accompanied by the repetitive current increase with lower amplitude in a stepwise manner is attributable to initiation of PbSCC composed of the local breakdown and repassivation of the surface oxide film. The potential drop with longer time interval accompanied by the current increase with larger amplitude is mainly due to the propagation of PbSCC. In addition, it was found that initiation of PbSCC in the medium mean free time period and its propagation in the long mean free time period were clearly distinguishable from uniform corrosion in the short mean free time period with different m values in the Weibull probability plots.

The EN technique, with the help of the stochastic approach using the Weibull distribution function, is one of the powerful tools for in-situ monitoring of crack initiation as the direct-current potential drop (DCPD) method for in-situ crack growth rate testing of crack propagation. To establish a reliable in-situ technique for monitoring of crack initiation, the reproducibility of EN signals and their relationship to the Weibull parameters for crack initiation should be carefully checked further under various conditions of primary and secondary water chemistries of PWRs.

Acknowledgement

This work was funded by the Korea Ministry of Education, Science and Technology.

References

1. H. R. Copson and S. W. Dean, *Corrosion,* 21 (1965), 1.
2. T. Sakai, S. Okabayashi, K. Aoki, K. Matsumoto and Y. Kishi, 'A study of oxide thin film of Alloy 600 in high temperature water containing lead', in *Corrosion/90*, Paper no. 520. NACE, Houston, TX, 1990.
3. S. S. Hwang, U. C. Kim and Y. S. Park, *J. Nucl. Mater.,* 246 (1997), 77.
4. S. S. Hwang, H. P. Kim, D. H. Lee, U. C. Kim and J. S. Kim, *J. Nucl. Mater.,* 275 (1999), 28.

5. J. Lumsden, A. McIlree, R. Eaker, R. Thomson and S. Slosnerick, 'Effects of Pb on SCC of Alloy 600 and Alloy 690 in prototypical steam generator chemistries', in Proc. 12th Int. Conf. on Environmental Degradation of Materials in Nuclear Power System – Water Reactors, ed. T. R. Allen, P. J. King and L. Nelson (Salt Lake City, UT, USA, 14–18 August 2005).

6. L. E. Thomas and S. M. Bruemmer, 'Observations and insights into Pb-assisted stress corrosion cracking of Alloy 600 steam generator tubes', in Proc. 12th Int. Conf. on Environmental Degradation of Materials in Nuclear Power System – Water Reactors, ed. T. R. Allen, P. J. King and L. Nelson (Salt Lake City, UT, USA, 14–18 August 2005).

7. H. P. Kim, S. S. Hwang, J. S. Kim and J. H. Hwang, 'Stress corrosion cracking of steam generator tubing materials in lead containing solution', in Proc. 13th Int. Conf. on Environmental Degradation of Materials in Nuclear Power Systems, ed. T. R. Allen and P. J. King (Whistler, Canada, 19–23 August 2007).

8. J. Stewart, D. B. Wells, P. M. Scott and D. E. Williams, *Corros. Sci.*, 33 (1992), 73.

9. R. A. Cottis, M. A. A. Al-Awadhi, H. Al-Mazeedi and S. Turgoose, *Electrochim. Acta*, 46 (2001), 3665.

10. R. A. Cottis, *Corrosion,* 57 (2001), 265.

11. H. A. A. Al-Mazeedi and R. A. Cottis, *Electrochim. Acta,* 49 (2004), 2787.

12. J. M. Sanchez-Amyay, R. A. Cottis and F. J. Botana, *Corros. Sci.,* 47 (2005), 3280.

13. T. Anita, M. G. Pujar, H. Shaikh, R. K. Dayal and H. S. Khatak, *Corros. Sci.,* 48 (2006), 2689.

14. K. H. Na, S. I. Pyun and H. P. Kim, *Corros. Sci.,* 49 (2007), 220.

15. K. H. Na and S. I. Pyun, *Corros. Sci.,* 49 (2007), 2663.

16. K. H. Na, S. I. Pyun and H. P. Kim, *J. Electrochem. Soc.,* 154 (2007), C349.

17. R. J. Jacko, *Corrosion Evaluation of Thermally Treated Alloy 600 Tubing in Primary and Faulted Secondary Water Environments*, EPRI NP-6721, Pittsburgh, PA, USA, 1990.

18. U. C. Kim, K. M. Kim and E. H. Lee, *J. Nucl. Mater.,* 341 (2005), 169.

19. T. Shibata, *Corros. Sci.,* 31 (1990), 413.

20. J. C. Uruchurtu and J. L. Dawson, *Corrosion,* 43 (1987), 19.

21. J. C. Uruchurtu, *Corrosion,* 47 (1991), 472.

22. T. Fukuda and T. Mizuno, *Corros. Sci.,* 38 (1996), 1085.

23. S. I. Pyun, E. J. Lee and C. H. Kim, *Surf. Coat. Technol.,* 62 (1993), 480.

24. S. I. Pyun, E. J. Lee and G. S. Han, *Thin Solid Films,* 239 (1994), 74.

25. J. J. Park and S. I. Pyun, *Corros. Sci.,* 46 (2004), 285.

26. T. Shibata, *Corros. Sci.,* 49 (2007), 20.

7

In-situ electrochemical impedance and noise measurements of corroding stainless steel in high-temperature water*

Jan Macák, Petr Sajdl, Pavel Kučera and Jan Vošta

Power Engineering Department, Institute of Chemical Technology Prague,
Technická 5, 166 28 Prague 6, Czech Republic

jan.macak@vscht.cz

Radek Novotný

Institute for Energy, Joint Research Centre, 1755ZG Petten, The Netherlands

7.1 Introduction

Growth of thermally formed oxides, dissolution of oxides and deposition of corrosion products play an important role in the performance of power cycles. Regarding the secondary side of PWR plants, corrosion concerns are typically associated with the integrity of the steam generator. Localised forms of corrosion in crevices and under deposits are considered to be potentially hazardous and are directly related to secondary circuit chemistry. The contribution of in-situ applied electrochemical techniques, especially corrosion potential and polarisation measurements, for corrosion studies in the light water reactor environment is covered in the comprehensive review of Turnbull and Psaila-Dombrowski [1]. AC techniques can provide additional information about the electronic properties and morphology of the oxide layer and about charge and mass transfer phenomena [2]. New techniques such as contact resistance (CR) or contact electrical impedance (CEI) [3] enable in-situ estimation of oxide resistivity, while controlled distance electrochemistry (CDE) and its derivative thin layer electrochemical impedance spectroscopy (TLEIS) allow for electrochemical measurements in high resistivity media such as boiling water reactor (BWR) coolant [4]. Electrochemical noise (EN) is another prospective electrochemical technique employed for characterisation of corrosion phenomena in the high-temperature environment. EN measurements were used, for example, to detect stress corrosion cracking initiation in BWR media [5].

The impact of impurities on the corrosion of steam generator (SG) tubes made from 08CH18N10T stainless steel was studied in-situ by cyclic voltammetry, CR and electrochemical impedance spectroscopy (EIS) measurements [6]. It was found that sulphate ions may behave more aggressively towards oxide film than chloride. In-situ

* Reprinted from *Electrochimica Acta*, 51(17), Jan Macák, Petr Sajdl, Pavel Kučera, Radek Novotný and Jan Vošta, 'In situ electrochemical impedance and noise measurements of corroding stainless steel in high temperature water', pp. 3566–77, © 2006, with permission from Elsevier.

EIS measurements were performed to study the formation of porous magnetite [2] (connected to the phenomenon known as 'denting'), or the effect of alumino-silicate deposits on corrosion of SG tubes [7].

The chemical composition and electronic structure of thermally formed oxides were studied by Montemor et al. [8] using ex-situ impedance measurements, photo-electrochemical measurements and Auger analysis. It was shown that chromium and iron oxides formed under high temperatures exhibit semi-conductive properties similar to those of the thin passive films formed at room temperature. The influence of the composition, structure and morphology of the oxide layer on stainless steel corrosion in high-temperature water is discussed in numerous studies [9–11]. Such oxide layers generally have a duplex character: their internal part grows in a topot-axial way and is formed by chromium rich spinel oxide. The external part grows in an epitaxial way, is less compact and its composition is similar to magnetite and its structure is predominantly formed by inverse spinel. It is generally accepted that the internal layer is responsible for slowing down the corrosion reaction.

In the presented work, in-situ electrochemical impedance spectroscopy and noise measurements were used to monitor corrosion behaviour during exposure of austenitic steel 08CH18N10T to high-temperature alkalised water in the presence and absence of chloride and to characterise the properties of the oxide layer formed.

7.2   Experimental

The experimental instrumentation used for in-situ measurements included an autoclave system and pressurised loop with facilities for temperature, pressure and chemistry control and a Gamry PC4/750 for electrochemical measurements and data acquisition (Fig. 7.1).

The autoclave body was constructed from Hastelloy C-276, and the fittings and autoclave lid bushings for the electrode system accommodation were made from 316 stainless steel. All EIS and polarisation measurements were performed in a three-electrode arrangement. The working electrode (WE) was machined from Ti-stabilised austenitic steel 08CH18N10T (Table 7.1). The working electrodes were tubular segments 1 cm high. Only their outer surface was exposed; each electrode thus had 5 cm^2 of exposed area. The WE holder was designed to accommodate two electrodes insulated from each other and from the holder body by PTFE spacers. The specimen surface was polished with a series of silicon carbide papers down to 1200 grit. Up to six identical samples in PTFE holders were separately placed in the autoclave and used in subsequent ex-situ measurements. The counter electrode (CE) was Pt wire wound coaxially around the WE. The end of the wire was anchored in a PTFE holder that was part of the bottom of the WE body. The external pressure-balanced Ag/AgCl electrode, filled with saturated KCl was used as the reference electrode (RE). To prevent leakage of chloride from the RE to the autoclave, the reference electrode bridge was equipped with two porous zirconia liquid junctions. The compartment between the junctions was filled with the autoclave electrolyte.

All reagents used were of analytical grade. Deionised water was prepared with a IWA20iol demi station and its conductivity was 0.055 μS cm^{-1}. Two electrolytes were used in the experiments: (i) deionised water with pH adjusted to 9.5 (at 25°C) with KOH and (ii) the same electrolyte but containing 200 ppm of chloride added as KCl.

The electrolyte was de-aerated by purging the storage vessel with argon and simultaneously circulating it through the autoclave. After the oxygen concentration

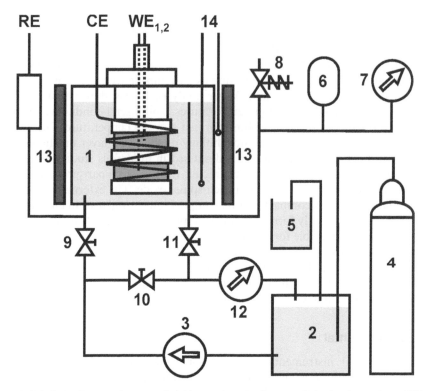

7.1 Autoclave and pressurised loop: 1, autoclave and electrode system; WE$_{1,2}$, working electrodes; CE, counter electrode; RE, reference electrode; 2, storage vessel (glass); 3, high-pressure pump; 4, pressure argon bottle; 5, water seal; 6, hydro accumulator; 7, pressure gauge; 8, safety valve; 9–11, high-pressure valves; 12, oxygen meter; 13, heating; 14, thermocouples

Table 7.1 Nominal composition of the steel 08CH18N10T (in wt. %)

steel	C	Mn	Si	P	Cr	Ni	Ti	Fe
08CH18N10T	< 0.08	< 2	< 1	< 0.045	17–19	9.5–12	> 5 x C	balance

dropped below 40 ppb, heating was started. When the required temperature and pressure were reached (280°C and 8 MPa) the measurements were started. All of the measurements were performed under stagnant conditions. To ensure a stable electrolyte level in the autoclave, a constant pressure level of 8 ± 0.2 MPa was maintained during the experiments by pumping a small amount of de-aerated electrolyte into the autoclave between the measurements. The oxygen concentration in the make-up water was kept between 5 and 40 ppb except during periods of time when higher concentrations were maintained to study the effect of oxygen on the electrochemical parameters. Each experiment was run for approximately 300 h, during which time 20–40 EIS and EN measurements were performed. EIS measurements were carried out at the open circuit potential (OCP) in the frequency range from 100 000 to 0.001 Hz at an amplitude of 5 mV. After a sufficient exposure time (usually about 40 h) sufficiently stable conditions were reached and EIS measurements could be run to very low frequencies (40–60 μHz).

EN measurements were also performed in a three-electrode arrangement – with the two identical working electrodes and the platinum wire as a reference electrode. Potential and current noise signals were measured simultaneously. Noise data were then processed by discrete Fourier transformation to identify the DC or slow oscillation part of the signal, corresponding to phenomena other than corrosion (e.g. heating cycles). The analysis of the noise data was then performed in the time domain and standard deviations of both signals were calculated.

At the end of the experiments, polarisation measurements were performed using a 100 mV/h scan rate. The polarisation data were post-run compensated for ohmic drop.

Alternatively, capacitance (or Mott/Schottky) measurements were performed at frequencies of 100, 500 or 1000 Hz. The polarisation was applied in 50 mV increments, starting from cathodic potentials, in the potential range ±1 V vs. corrosion potential. At each potential, the impedance was measured successively at the given frequencies. The polarisation time at each potential was as short as possible (max. 12 s) to keep the electric charge low and to minimise possible chemical changes in the oxide due to the applied polarisation.

7.3   Results and discussion

7.3.1   Electrochemical noise measurements

In the presented work, EN data were used to calculate noise resistance R_n. The equivalence of noise resistance and polarisation resistance, R_p, was confirmed by the work of Chen and Bogaerts [12]. Therefore, the noise resistance represents a corrosion parameter which could be compared directly with corrosion data obtained from impedance measurements, namely with polarisation resistance. Noise resistance was calculated as the ratio of standard deviations of the simultaneously measured potential and current noise signals:

$$R_n = \frac{\sigma(V)}{\sigma(I)} \qquad [7.1]$$

In Fig. 7.2, the noise resistance and polarisation resistance values obtained from impedance measurements are compared. Clearly, good agreement was reached between the two values for both of the cases (the presence and absence of chloride).

7.3.2   Impedance measurements

The impedance spectra for a system without chloride are shown in Fig. 7.3. From the Nyquist plot (Fig. 7.3a), it is evident that the general shape of the spectra includes the impedance response of a corrosion process at higher frequencies and a mass transfer process at lower frequencies. Such a shape of impedance dispersion can be characterised by a $-R_e[(R_pZ_w)CPE_{dl}]$-equivalent circuit (Fig. 7.4a), where R_e is the solution resistance, the serial combination R_pZ_w corresponds to the faradaic electrode process and includes the polarisation resistance R_p and the Warburg impedance Z_w associated with semi-infinite diffusion. CPE_{dl} is connected in parallel to R_pZ_w and represents a part of non-faradaic electrode phenomena – namely the response of a non-ideal (dispersive) double layer capacitance. CPE is a common abbreviation for

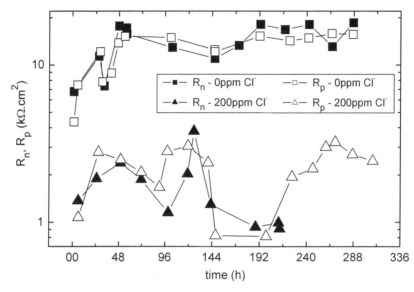

7.2 Time dependence of noise resistance R_n and polarisation resistance R_p estimated from EIS data

the constant phase element, the impedance of which is expressed as $Z_{CPE} = Q^{-1}(j\omega)^{-n}$ where Q is the CPE coefficient, n the CPE exponent, j the imaginary unit and ω the angular frequency. The CPE exponent n is a measure of the capacitance dispersion having values between 1 (ideal capacitance) and 0.5 (highly dispersed capacitance, e.g. at porous electrodes).

The other part of electrode capacitance is represented by the response of the oxide layer. It does not form a distinct shape in Nyquist or Bode plots and cannot be derived from overall impedance spectra fitting. The method used for its estimation will be described in the next subsection.

In the presence of chloride ions (Fig. 7.5), a slightly different picture can be seen. From the Bode diagram in Fig. 7.5b, it is evident that another capacitive time constant appears in the mid-frequency range (1–100 Hz). Because this time constant is linked to the presence of chloride, it is likely related to the adsorption of surface active chloride anions on the electrode surface. Therefore, the experimental imped-ance spectra were fitted by the equivalent circuit containing two additional elements – $R_e(R_aCPE_a)[(R_pZ_w)CPE_{dl}]$ (Fig. 7.4b), where R_a and CPE_a are adsorption resistance and adsorption dispersive capacitance, respectively. The resistance R_a remained almost unchanged over the whole exposure time and had relatively low values between 50 and 150 Ω cm^2 compared to the polarisation resistance (in the case of Cl$^-$ presence usually 2000 to 4000 Ω cm^2). Such low resistance values could corre-spond to a local corrosion process proceeding at a very high rate (e.g. active pitting corrosion). A careful ex-situ examination of the electrode surface after the experi-ment did not show any local corrosion attack during the experiment. An adsorption impedance of a similar type has not been documented yet on steel or oxide surfaces but has often been observed in the presence of specifically adsorbed anions on various metals [13,14]. However, a detailed discussion of R_a and CPE_a is beyond the scope of this paper.

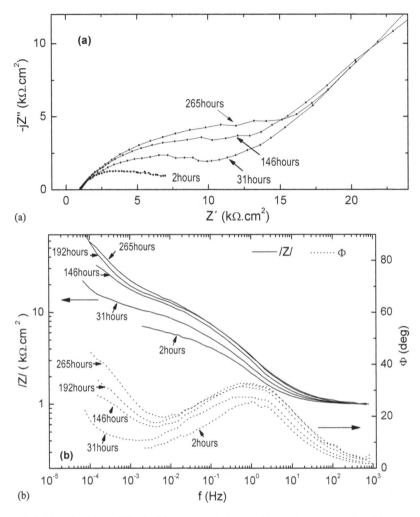

7.3 Nyquist (a) and Bode (b) representations of impedance spectra. Measured in alkalised water at OCP. All of the spectra in Bode diagram are normalised using $R_e=1\ k\Omega$

The frequency dispersion of impedance can also be described by a complex capacitance vs. frequency plot. It should be mentioned that, in the presented case, the transformation of impedance into 'capacitance' should only be considered as a formal tool because other impedance contributions are also present in the experimental spectra. The complex capacitance is related to the impedance as $C^*(\omega) = [j\omega(Z^*(\omega)-Z(\omega)]^{-1}$, where j and ω have the usual meaning, $Z^*(\omega)$ is the complex impedance and $Z(\omega)$ is the high frequency limit of the electrode impedance, which is usually equivalent to the solution resistance. In the complex 'capacitance' plot shown in Fig. 7.6, four distinct areas can be recognised. The high frequency part of the spectrum is dominated down to the ~5 kHz level by a very low capacitance (~500 pF), which was practically ideal (not-dispersive) and remained constant during the whole exposure. This is apparently a measurement artefact resulting from the responses of the leads, liquid junction, counter electrode arrangement, etc. The oxide capacitance

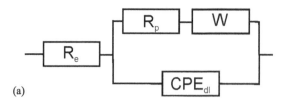

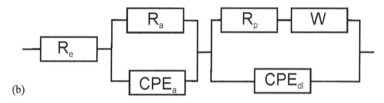

7.4 Equivalent circuits used for the data fitting: $R_e[(R_pZ_w)CPE_{dl}]$ (a) and $R_e(R_aCPE_a)[(R_pZ_w)CPE_{dl}]$ (b)

is hidden in the flat part of the capacitance spectrum between ~10 kHz and 100 Hz. Its occurrence in the higher frequency domain is probably a consequence of its low values and its association with an oxide (or space charge layer) resistance. The resulting time constant is rather low due to the small magnitudes of both elements. In the absence of chloride, the oxide response extends down to <10 Hz. Another region (~10 mHz to ~100 Hz) corresponds to the response of the double layer formed at the oxide–electrolyte interface. In the presence of chloride, the adsorption capacitance is dispersed in a very similar way to the double layer capacitance and cannot be clearly separated from it in the complex capacitance plot (the values of the exponents n of the corresponding CPE elements are very close).

The low frequency range (below ~10 mHz) is dominated by the Warburg impedance that is associated with mass transfer phenomena. A more detailed discussion of the single impedance contributions follows in the next subsections.

Impedance of oxide layer

Oxide layer capacitances are generally very low and the impedance response of oxide layers is usually located in the high frequency domain of the impedance spectra. In this study, the oxide layer impedance spreads over 2–3 decades of the frequency range (10^3–10^1 Hz or 10^4–10^2 Hz for the absence or presence of chloride, respectively) and is limited from the high frequency side by the parasitic impedances and from the low frequency side by the impedance of the double layer.

For high and medium frequency domains, where capacitive contributions to the electrode impedance are dominant, the complex impedance $Z^*(\omega) = Z' - jZ''$ can be taken as equal to a serial combination of complex capacitance $C^*(\omega) = C' - jC''$ and pure resistance R_e (which includes electrolyte resistance and parasitic resistance contributions). In such a case

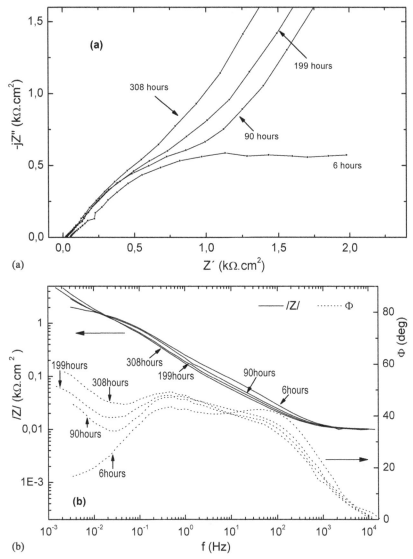

7.5 Nyquist (a) and Bode (b) representations of impedance. Measured in alkalised water, in the presence of 200 ppm Cl⁻. The spectra in Bode diagram are normalised using $R_e = 10\ \Omega$

$$C' = \frac{1}{\omega} \frac{Z''}{(Z' - R_e)^2 + Z''^2} \qquad [7.2]$$

and

$$C'' = \frac{1}{\omega} \frac{Z' - R_e}{(Z' - R_e)^2 + Z''^2} \qquad [7.3]$$

From equations 7.2 and 7.3, it follows that the R_e value must be estimated very carefully as it significantly affects the shape of the high frequency part of the complex

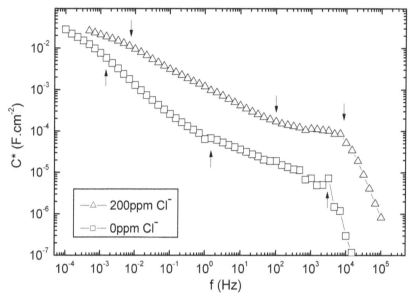

7.6 Bode representation of complex capacitance, measured after 265 h (chloride absence) and 260 h (chloride presence) of exposure. The arrows mark the breakpoints of different capacitance dispersion domains

capacitance plot [15]. Figure 7.7 shows a part of the complex capacitance plot (Cole–Cole plot) obtained after 30 h in the system without chloride. The cluster of points at the highest frequencies ($>~$kHz) corresponds to the parasitic high frequency end of the impedance spectrum. The linear part, which extends over almost 3 frequency decades is associated with the dielectric dispersion of the oxide film and can be described by a universal law of the dielectric response [16,17]

$$C*(\omega) = C_{\infty} + \frac{\Delta C}{(j\omega)^n} \qquad [7.4]$$

where $C*(\omega)$ is complex capacitance, ΔC is a constant associated with dielectric dispersion, the exponent n represents the extent of dielectric dispersion and corresponds to the angle Φ between the real axis and the experimental curve, $\Phi = n.\ 90°$, and C_{∞} is a parameter estimated from the high frequency intercept of the linearly extrapolated experimental curve and the real axis and meaning pure, non-dispersed capacitance. The second term in equation 7.4 can be considered as a form of the CPE, hence equation 7.4 expresses the capacitance of a parallel combination of an ideal capacitance C_{∞} and CPE. This concept was used to analyse data obtained in studies of corrosion layers on zirconium alloys [18–20]. In the case of zirconium oxide, which behaves more or less as an insulating dielectric, the C_{∞} value can be related directly to the total oxide layer thickness. However, oxides formed on stainless steels are known to be n-type semiconductors and homogeneous dielectric behaviour of the entire oxide layer cannot be expected. Instead, the dielectric response to a potential perturbation is usually localised in the external part of the oxide at the oxide/electrolyte interface, which is called the space charge layer. Applying the relationship for a parallel plate

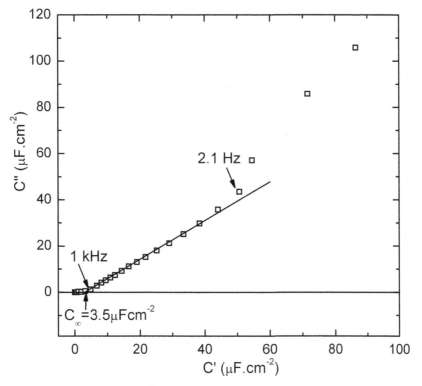

7.7 Cole–Cole diagram of steel electrode capacitance, measured in alkalised water at 280°C after 30 h of exposure

capacitor, one can relate the values of C_∞ directly to the thickness of the space charge layer

$$d_{\mathrm{W}} = \frac{\varepsilon_{\mathrm{r}}\varepsilon_{\mathrm{o}} r A}{C_\infty} \qquad [7.5]$$

where d_{W} is the thickness of the space charge layer, ε_{o} is the vacuum permittivity, ε_{r} is the relative permittivity of the oxide (dielectric constant), r is the roughness factor and A is the electrode geometric area.

Values of ε_{r} for oxides formed on stainless steels reported in the literature [21–24] are generally between 10 and 30. For thermally grown oxides and oxides formed in hot water, $\varepsilon_{\mathrm{r}} = 12$ is often used [8,24] and this value is also used in this study. When higher values ($\varepsilon_{\mathrm{r}} = 40$) are considered [25], usually the effect of soaking the oxide in water at ambient temperature is taken into account. Water at 280°C and 8 MPa has $\varepsilon_{\mathrm{r}} = 22.9$, therefore its presence in the oxide pores will affect the overall dielectric constant much less than soaking at ambient temperature [26] ($\varepsilon_{\mathrm{r}} = 80$ at 20°C).

A more significant effect can be expected from the surface roughness, which is likely to increase during the experiment due to external oxide formation. Roughness factors were obtained from ex-situ double layer capacitance measurements performed in borate buffer. The C_{dl} value of highly polished steel samples (diamond paste grit, 0.1 μm) was about 20 μF cm^{-2} while C_{dl} of experimental samples was between 80 and

$180 \ \mu F \ cm^{-2}$, which gives a roughness factor between 4 (Cl⁻ absence, shorter exposure in autoclave) and 9 (Cl⁻ presence, longer exposure). For given r values and values of C_∞ presented in Fig. 7.8, d_W values of about 12–21 nm were obtained for oxide formed in the absence of chloride and 3–5 nm in the presence of chloride. Comparing these numbers with the total oxide thickness of 2.7–3.2 μm estimated by SEM, the space charge layer occupies less than 0.7% of it. From Fig. 7.8, it is also evident that the value of C_∞ is almost constant during the whole exposure time and close to the value obtained under the same conditions before heating was started. On the other hand, in the presence of chloride, a large increase in C_∞ was seen after heating was started and the C_∞ values correspond to the oxygen ingress to the autoclave.

The thickness of the space charge layer is related to the density of dopants (in our case donors) in the oxide. The potential distribution across the space charge layer can be expressed as [27]

$$\phi(sc) = -\left(\frac{e_o N}{2\varepsilon}\right) d_W^{\ 2} \qquad [7.6]$$

where e_o is the electronic charge, ε is the dielectric constant and N is doping density. For a slightly doped oxide, the magnitude of d_W must be larger to provide for the same $\phi(sc)$. The doping density is equal to the charge carrier concentration in oxide, which plays a crucial role in oxide growth models [28,29] and together with other electronic properties, is correlated with oxide/passive film protectiveness [30,31]. Thus one could conclude that the oxide formed in the presence of Cl⁻ has a higher concentration of defects, which would mean worsening of its protective properties.

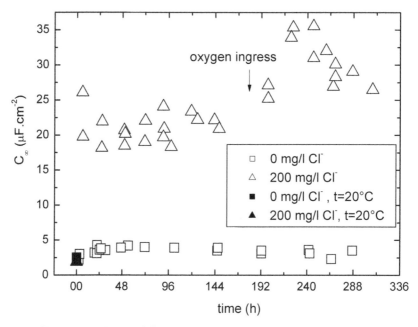

7.8 Time dependence of C_∞ capacitance, measured in alkalised water at 280°C. Full points – measured at ambient temperature

Unfortunately, there are some indications showing that correlating C_∞ values estimated in-situ in the presence of chloride with d_W could be incorrect. First, the magnitude of C_∞ rises immediately with oxygen ingress and thus with the corrosion rate (Fig. 7.8) and this rise is reversible. Second, the in-situ measured values do not match the ex-situ values measured in borate buffer [32] (the former being 18–34 μF cm^{-2}, while the latter is 3–4 μF cm^{-2}). Both facts indicate that the real reason for the increase in in-situ measured capacitance is not connected with the bulk oxide electronic properties. The in-situ measured C^{-2} vs E plot (or Mott–Schottky plot) presented in Fig. 7.9 together with the polarisation curve gives the same indication. The doping density can be estimated from the linear part of the plot according to the Mott–Schottky equation

$$\frac{1}{C_{sc}^2} = \frac{2}{\varepsilon_0 \varepsilon_r e_o r^2 N} \left(|E - E_{fb}| - \frac{kT}{e_o} \right) \qquad [7.7]$$

where C_{sc} is the space charge layer capacitance, E the applied potential, E_{fb} the flat-band potential, k the Boltzmann constant, and T the temperature. The Boltzmann term kT/e_0 is equal to 0.046 V at 280°C. The shape of the curve is very similar to the ex-situ measured Mott–Schottky plots of 08CH18N10T steel [32] or 316L steel [8] pre-exposed in an autoclave in alkalised water but there are two significant differences – values of capacitance are more than one order of magnitude higher and the curve is shifted to more positive potential values.

Two marked changes of slope can be observed in the Mott–Schottky plot (Fig. 7.9): the first one at $E \approx -0.7$ V and the second one at $E \approx -0.1$ V (vs. Ag/AgCl). Both potentials coincide with potentials of the current increase in the polarisation curve and correspond approximately to the high-temperature equilibrium potentials

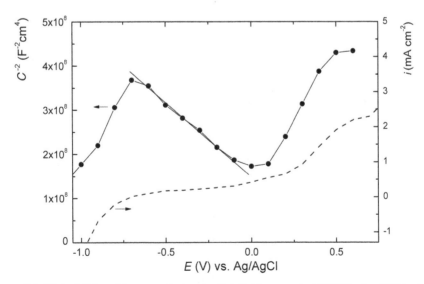

7.9 Plot of C^{-2} and i vs. applied potential E. Measured at 280°C after 307 h of exposure in alkalised water in the presence of 200 ppm Cl$^-$

for Fe(III)/Fe(II) and Cr(VI)/Cr$_2$O$_3$ couples [33,34]. Fe(II) species formed at cathodic potentials and Cr(VI) formed at anodic potentials can be partially incorporated in the oxide and affect its electronic properties by functioning as the doping species. Because the potential scan was started at cathodic potentials, Fe(II) species formed may contribute to the high doping density $N = 1.7 \times 10^{22}$ cm^{-3} which was estimated from the slope of the Mott–Schottky plot in the potential range $-0.7 < E < -0.1$ V.

The high magnitude of N corresponds to high C_∞ values and exceeds values obtained in ex-situ experiments [8,31] by more than two orders of magnitude. However, the contribution of the Fe(II) species formed at negative potentials could not be the sole reason for very high N and capacitance values, because comparable C_∞ values were observed in experiments performed at OCP. Another reason for high N values may be associated with the surface phenomena rather than bulk oxide properties, especially with physico-chemical processes at the oxide/electrolyte and oxide/metal interfaces. Both oxide dissolution and metal oxidation in the presence of chloride are accelerated by increased concentration of dissolved oxygen. The oxide dissolution leads to a higher density of surface heterogeneities (cationic vacancies). Both cationic vacancies and chloride ions adsorbed on the oxide surface [36], may cause generation of surface states at the oxide/electrolyte interface. Surface states on the semiconductor surface take part in charge transfer and cause an excess capacitance. That is often a reason why capacitance measurements cannot be used in the determination of the bulk electronic properties of semiconductors [34,35]. Surface states were also reported to cause a notable deviation from linearity of Mott–Schottky plots of iron oxide passive films due to the increase in oxide capacitance [30]. The surface state density can be calculated according to the equation [36]

$$e_o N_{ss} = q_{ss} = \int_{E_0}^{E_1} C_{ss} dE \qquad [7.8]$$

where N_{ss} is the surface states density, e_o the electronic charge, q_{ss} the charge density associated with surface states, C_{ss} the surface state capacitance and E is potential.

In our case it is not exactly clear where the relevant potential range E_0–E_1 lies; integration, if performed between -0.7 V and -0.1 V vs Ag/AgCl, gives $N_{ss} \approx 2 \cdot 10^{14}$ cm^{-2}. Just for comparison, the calculated N_{ss} magnitude is almost 50 times higher than for surface states on anodically formed iron oxide [30]. Such a high value, if interpreted as surface oxide cationic vacancies, can still have physical meaning as it is certainly lower than the surface density of iron atom sites of external oxide (e.g. surface density of Fe sites on ideal (001) magnetite plane is approximately $2.2 \cdot 10^{14}$ cm^{-2}).

Polarisation resistance and double layer capacitance

From the corrosion standpoint, the most important contribution to the overall electrode impedance is the polarisation resistance $R_p = (dE/di)_{i=0}$. Its relationship to the corrosion current has the following form for the case of the diffusion controlled cathodic reaction [37]

$$i_{corr} = \frac{\beta_a}{2.303 R_p} = \frac{B}{R_p} \qquad [7.9]$$

where β_a is the Tafel parameter for the anodic reaction.

The polarisation resistance values obtained by analysing impedance spectra are presented in Fig. 7.2. Both in the absence and presence of chloride, a steep increase in R_p values can be seen at the beginning of the experiment that corresponds to the formation of a dense and homogenous oxide layer, which is able to slow down the corrosion rate. After reaching a certain exposure time (30–70 h), the increase in polarisation resistance in the absence of chloride is rather slow (13.8 kΩ cm^2 at 48 h to 15.9 kΩ cm^2 at 288 h). The corrosion behaviour was found to be markedly dependent on the oxygen concentration and the local maxima and minima in the R_p profile were associated with variation of the residual dissolved oxygen concentration in the feed water.

The variation of R_p with time is more pronounced in the presence of chloride. The deep local minimum of R_p at ~190 h (Fig. 7.10) corresponds to the peak ingress of oxygen during the water pumping. From the oxygen concentration in the feed water (272 ppb) and the amount of water pumped in the autoclave, it can be concluded that the amount of oxygen was ~15 µg and the increase in the autoclave oxygen concentration was ~50 ppb. The oxygen ingress also influenced other electrochemical parameters. There was a steep increase in the corrosion potential profile and a gradual increase in conductance R_e^{-1} with time. As the geometry of the electrode system and the temperature remained constant throughout the experiment, conductance changes have to be related to the variation in the solution composition. The conductance increase over time is caused by the increase in concentration of soluble corrosion products in the autoclave. It can be seen that the R_e^{-1} increase is steeper after the oxygen ingress and continues until the residual oxygen in the autoclave has been consumed. In the final phase of the experiment, R_e^{-1} decreases as a consequence of the drop in solubility of iron corrosion products (oxide deposition reaction prevails over oxide dissolution).

A relatively high solubility of iron in alkalised high-temperature water in the presence of chloride was confirmed by inductively coupled plasma–optical emission spectrometry (ICP-OES) at the end of the experiments. The results presented in

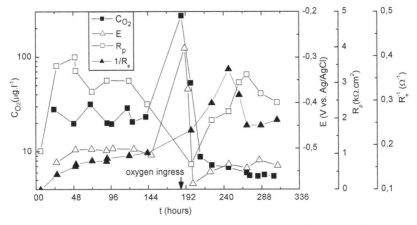

7.10 Influence of the increase in oxygen concentration C_{O_2} in autoclave feed water on electrode potential E, polarisation resistance R_p and solution conductivity R_e^{-1}. The arrow denotes the point where $C_{O_2} = 272$ µg l^{-1}

Table 7.2 Oxide thickness estimated by SEM (mean values in brackets) and soluble corrosion product concentrations obtained by ICP-OES analysis of autoclave content after 300 h of exposure at 280°C. In the case of Fe, the concentrations are the results of three independent analyses.

Cl⁻ content	oxide thickness	soluble corrosion products		
	μm	Fe μg/l	Cr* μg/l	Ni** μg/l
0 ppm	3.2–3.4 (3.3)	20±3	<5	<10
200 ppm	2.1–2.9 (2.7)	1500±300	<5	14

* – below the detection limit
** – possible interference from internal autoclave surfaces

Table 7.2 also indicate a very low solubility of iron in pure alkalised high-temperature water which is in good agreement with literature data [11,38]. The concentration of iron in the autoclave solution was only ~20 ppb, contrary to its high solubility in the system with chloride (~1500 ppb). The solubility of chromium and nickel remains very low in both cases.

Regarding the low solubility of corrosion products in pure alkalised water, the deposition of iron oxides on the autoclave surfaces should also be relatively insignificant. It should then be possible to obtain at least a rough correlation between the corrosion rate estimated from electrochemical data (instantaneous rate) and from oxide thickness (integral rate). It was assumed that the oxide is compact magnetite of specific density equal to 5.2 g cm⁻³. The average oxide thickness estimated by ex-situ scanning electron microscopy (SEM) performed on cross-section samples was 3.3 μm which gives 1.72 mg cm⁻² of overall mass of magnetite formed per unit area. It corresponds to 1.24 mg of reacted Fe per cm². From Faraday's law, a rough estimate of the charge density passed due to corrosion (oxide formation) can be made

$$q = i_{corr} t = \frac{zFm}{M} \qquad [7.10]$$

where q is the charge density, i_{corr} the corrosion current density, t the exposure time ($t = 290$ h $= 1.044 \cdot 10^6$ s), z the number of electrons ($z = 2.667$ for oxidation of Fe to magnetite), F the Faraday's charge (96 484 C mol⁻¹), and M the molar mass of Fe (55.8 g mol⁻¹). Using $m = 1.24 \cdot 10^{-3}$ g cm⁻², the corrosion charge density $q = 5.5$ C cm⁻² is obtained, which corresponds to an average corrosion current density $i_{corr} = 5.3$ μA cm⁻². As stoichiometric, compact magnetite was assumed, which is not true especially for the external part of the oxide, it can be concluded that the above value of corrosion charge represents the upper limit and the real value should be lower.

To convert the instantaneous corrosion parameters to the integral ones, the mean integral value of polarisation resistance R_i is introduced as

$$R_i = \frac{\int_0^t R_p dt}{t} \qquad [7.11]$$

where R_p is the polarisation resistance and t the total exposure time. The steady corrosion current density can then be calculated using equation 7.9. The B value of 37 mV calculated from literature data [39] can probably also be used in our case. The

B values obtained from our polarisation measurements performed after 300 h exposure on electrodes covered with thick oxide layers are higher, $B \approx 70$ mV [31]. The difference is probably caused by the modification of the electrode surface, which has, in the second case, much lower corrosion activity. For *B* values 37 mV $\leq B \leq$ 70 mV, an average corrosion current between 2.6 and 5.1 μA cm^{-2} can then be obtained. This corresponds to the mean corrosion charge density values $2.7 \leq q \leq 5.3$ C cm^{-2}. It can be seen that the upper charge density limit of 5.5 C cm^{-2} calculated from the compact oxide layer thickness was not exceeded.

In the presence of chloride, the iron solubility was much higher. As the dissolved iron can readily scale on the internal autoclave surfaces, it was not possible to make any comparison between instantaneous and integral corrosion rates.

Together with polarisation resistance, double layer capacitance forms a time constant in the middle and low frequency domain between approximately 10^1 and 10^{-1} Hz. The capacitance values were estimated by fitting the experimental data with the impedance of equivalent circuits described in section 7.3.2. Double layer capacitance exhibited a highly dispersive character and was approximated by the CPE. The results are presented in Fig. 7.11. The value of CPE coefficient Q_{dl} was between 1.7 and $2 \cdot 10^{-4} \, \Omega^{-1}$ s^n cm^{-2} during the first 2 days, then decreased slowly to approximately $1.4 \cdot 10^{-4} \, \Omega^{-1}$ s^n cm^{-2} and remained almost unchanged until the end of the experiment.

A similar time profile can also be seen in the case of CPE$_{dl}$ exponent n_{dl}. The magnitude of n_{dl} indicates a relatively high degree of capacitance dispersion on the surface ($0.59 \leq n \leq 0.67$), an increase of which occurred during the first 48 h, most probably due to the formation of an external oxide layer having a higher degree of surface roughness than the original steel surface. Normalisation of CPE values to the 'loss capacitance' can be performed using the following formula [40]

$$Z_{CPE_{dl}} = V^{-1} \omega_o^{-1} \left(\frac{j\omega}{\omega_o} \right)^{-n} \qquad [7.12]$$

where Z_{CPE} is CPE impedance, *V* is loss capacitance and $\omega_0 = 2\pi f_0$ is the normalisation frequency, which should be chosen from a frequency range where the given capacitance has the highest significance. If normalised to $f_0 = 1$ Hz, a double layer capacitance in the range of 60 to 100 μF cm^{-2} was obtained, which is relatively close to the values measured in the studied system at ambient temperature.

Similar data for the ingress of oxygen in the presence of chloride during the experiment show more dramatic evolution in time (Fig. 7.11b). It is evident that Q_{dl} values are much higher in this case and that both Q_{dl} and n_{dl} respond distinctly to the ingress of oxygen described above. The steady value of Q_{dl} reaches approximately $1.5 \cdot 10^{-3} \, \Omega^{-1}$ s^n cm^{-2} and is one order of magnitude higher than in the absence of chloride. During a period of increased corrosion (the early stages of the experiment and following the ingress of oxygen), its value almost doubled (up to $2.5 \cdot 10^{-3} \, \Omega^{-1}$ s^n cm^{-2}). The corresponding loss capacitance (again normalised to $f_0 = 1$ Hz) was ~900 μF cm^{-2} during the steady state and rose to 1500 μF cm^{-2} when the corrosion rate increased. A significant decrease in the exponent n_{dl} occurred in the final part of the experiment when oxygen was consumed and the balance between oxide/steel dissolution and oxide deposition was shifted in favour of deposition. The dissolved iron is deposited as external oxide crystallites and this process leads to an increase in the surface roughness and thus capacitance dispersion. The values of capacitance are too high to be considered as a mere double layer capacitance – if the prevailing part of the

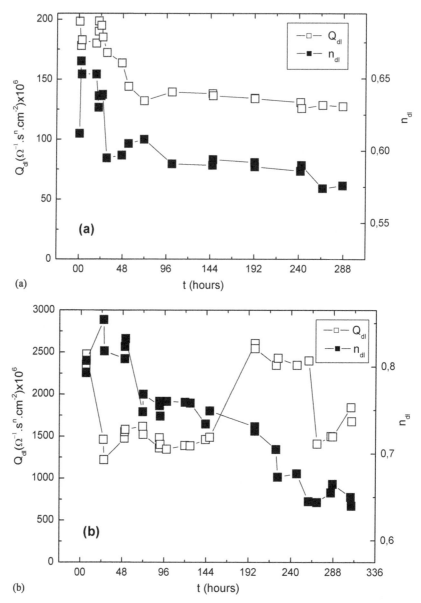

7.11 Time dependence of CPE_{dl} parameters. Measured in alkalised water at 280°C in the absence (a) and presence (b) of 200 ppm of chloride

capacitance increase had occurred in the compact (Helmholtz) layer, its thickness would have become too small to be reasonable ($H \approx 10^{-11}$ m). The contribution of increased ionic concentration to the diffusion (or Gouy–Chapman) layer capacitance can be quantified using the following relationship [41]

$$C_{G} = \left[\frac{F^{2}\varepsilon_{o}\varepsilon_{r}}{RT} \sum_{i} z_{i}^{2} c_{i} \right]^{1/2} \qquad [7.13]$$

where C_G is the capacitance of the Gouy–Chapman layer, z_i the charge of an ion i, c_i its concentration and other symbols have their usual meaning. For the potassium chloride added, we can estimate the sum $\Sigma z_i^2 c_i \approx 1.1 \cdot 10^{-5}$ mol cm^{-3} and thus $C_G \approx$ 7 μF cm^{-2}. Therefore, it can be seen that the contribution of C_G is relatively unimportant and the most probable reason for the capacitance increase in the presence of chloride is the coupling between the faradaic corrosion process and the non-faradaic interface charging process and the resulting faradaic pseudocapacitance, which is often associated with surface relaxation phenomena. Pseudocapacitance and double layer capacitance contributions to electrode impedance clearly overlap which makes evaluation of C_{dl} impossible. It is worth noting that a significant increase in capacitance also causes the overlap with the diffusion time constant and may cause problems in obtaining a reasonable magnitude of polarisation resistance [31,42].

Diffusion impedance

A mass transfer process dominates the electrode impedance below 10^{-1} Hz. The corresponding impedance dispersion has the form of a half-line and can be expressed as semi-infinite Warburg diffusion impedance connected with diffusion of dissolved oxygen from the bulk to the electrode surface

$$Z_W = \frac{\sigma}{\sqrt{\omega}} - j\frac{\sigma}{\sqrt{\omega}} = \frac{\sigma}{\sqrt{\omega}}(1-j) = \frac{\sigma\sqrt{2}}{\sqrt{j\omega}} \qquad [7.14]$$

where Z_W is the Warburg diffusion impedance and σ is the Warburg coefficient which in the case of a surplus of one component in a redox system is often presented as

$$\sigma = \frac{RT}{z^2 F^2 A \sqrt{2} c \sqrt{D}} \qquad [7.15]$$

where c is the bulk concentration of a reactive species, and D its diffusion coefficient. From equation 7.14, it follows that the type of Z_W dispersion is formally identical to that of Z_{CPE} for $n = 0.5$ and the Warburg coefficient can be related to the CPE coefficient

$$\sigma = \frac{1}{Q_W \sqrt{2}} \qquad [7.16]$$

where Q_W is the coefficient of the constant phase element used for the description of diffusion impedance and formally substituting Z_W in equivalent circuits used for data fitting. In alkalised water, n_W is close to the theoretical value of 0.5 for the major part of the experiment. Its gradual increase in time up to ~0.6 was observed after longer exposure times (200 h) and should be ascribed to the surface modification due to external oxide deposition. The same reason also applies for the increase in n_W values obtained in the presence of chloride. The n_W values are generally higher in this case (0.55–0.78) and their increase after long exposure time is steeper due to the greatly enhanced dissolution/deposition reaction and consequent increase in surface roughness. Similar changes in the slope of the Warburg line were proved to be linked to fractal geometry effects of the electrode surface [43,44].

The semi-infinite diffusion is a consequence of static conditions in the autoclave. The thermal convection was limited because the autoclave walls were heated and its lid and bottom were carefully insulated which minimised the thermal gradient. After the surface oxide was formed and the corrosion rate stabilised, it was possible to measure the impedance down to very low frequencies (e.g. $4 \cdot 10^{-5}$ Hz). However, due to very long measurement times, the reliability of the data at the lowest frequencies was rather limited and σ estimated from equation 7.16 usually depended on the value of the low frequency limit set for the impedance data fitting. In such a case, σ was determined more accurately from the Warburg plot as the slope of the linear part of $Z'' = f(\omega^{-0.5})$ dependence (Fig. 7.12). The linearity was usually observed at higher frequencies of the relevant frequency range. The deviation from linearity in the lowest frequency range was related to the drop in oxygen concentration due to its consumption during the measurement. Sometimes only an estimate of an instant σ value from a tangent to the experimental curve could be performed. A rough estimate can also be obtained from the low frequency limit (Fig. 7.12) taking into account that the drop in the oxygen concentration is relatively slow and will affect the Z'' values at higher frequencies only negligibly.

To evaluate oxygen concentration in the autoclave, the diffusion coefficient D is needed. As no literature data on oxygen diffusion coefficient in water at 280°C were available, they were estimated from the Einstein–Stokes equation [45]

$$D = \frac{RT}{6\pi\eta N_A r_{O_2}} \qquad [7.17]$$

where R is the gas constant, T the temperature, η the dynamic viscosity of water at 280°C ($9.4 \cdot 10^{-5}$ Pa s [24]), N_A is Avogadro's number, and r_{O_2} the molecular diameter of oxygen (here $1.09 \cdot 10^{-10}$ m was used). The calculated value of D was approximately $4 \cdot 10^{-4}$ cm² s⁻¹. Using this value in equation 7.15, the oxygen concentration in the

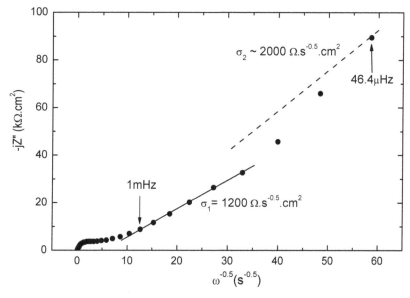

7.12 Warburg plot of the imaginary part Z''. Measured in alkalised water, $t = 280$°C, exposure time 192 h

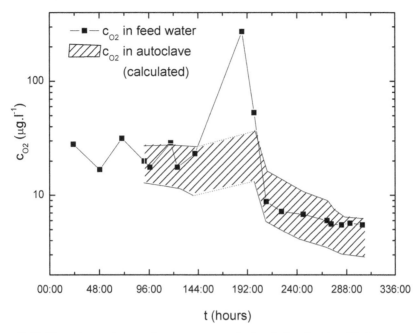

7.13 Time dependence of oxygen concentration in feed water, Warburg coefficient and calculated oxygen concentration in autoclave, 200 ppm of chloride

autoclave was found to be between 5 and 30 ppb. These values are reasonable considering the usual oxygen concentration in the feed water (5–40 ppb). In the specific case shown in Fig. 7.12, the oxygen concentration drop from ~30 ppb ($\sigma = 1200 \ \Omega \ s^{-0.5} \cdot cm^2$) to ~15 ppb (from a low frequency estimate of $\sigma = 2000 \ \Omega \ s^{-0.5} \cdot cm^2$) during about 18 h of impedance measurement can be estimated. Figure 7.13 shows a comparison between the oxygen concentration in the feed water and the range of values calculated from the Warburg coefficient range. Unfortunately, the estimate of σ immediately after the ingress of oxygen was not possible because the conditions were not sufficiently steady to allow measurements at low frequencies.

7.4   Conclusions

The following are the main conclusions from the experiments performed in high-temperature water

1. A good agreement was obtained between the polarisation resistance values estimated from impedance data and noise resistance.
2. The instantaneous corrosion rate obtained from electrochemical measurements correlates well with the integral corrosion rate estimated from the oxide thickness.
3. The application of the universal law of dielectric response allows the separation of oxide layer and double layer capacitances.
4. The dielectric thickness of the oxide estimated from the oxide capacitance is approximately 0.7% of the total oxide thickness and was interpreted as the

space charge layer thickness. The observed space charge layer capacitance values were very stable during the experiment in pure alkalised water. Consequently, no significant changes in the charge carrier concentration in the oxide layer could be expected.

5. In the case of increased corrosion and oxide dissolution rates in the presence of chloride, it was not possible to associate the oxide impedance response with bulk oxide electronic properties. A large increase in the oxide layer capacitance was attributed mostly to the effect of surface states formed due to the oxide dissolution.

6. A similar effect of the presence of chloride was seen in the case of double layer capacitance. Its significant increase was attributed to the faradaic pseudocapacitance effect associated with a higher dissolution rate.

7. All impedance parameters respond sensitively to the changes in oxygen concentration and therefore also to the changes in corrosion rate.

8. The observed semi-infinite type diffusion impedance was associated with dissolved oxygen molecular diffusion. The calculated range of autoclave oxygen concentrations correlates with feed water concentrations.

List of symbols

A	geometric area (cm^2)
B	constant, equation 7.9 (V)
c, c_i	concentration (mol l^{-1}, mol cm^{-3})
C_{dl}	double layer capacitance (μF cm^{-2})
C_∞	high frequency capacitance limit, equation 7.4 (μF cm^{-2})
C'	real part of capacitance (μF cm^{-2})
C''	imaginary part of capacitance (μF cm^{-2})
C^*	complex capacitance (μF cm^{-2})
CPE	constant phase element
CPE$_a$	constant phase element describing dispersive adsorption capacitance
CPE$_{dl}$	constant phase element describing dispersive double layer capacitance
CPE$_W$	constant phase element describing non-ideal diffusion impedance
C_G	capacitance of Gouy–Chapman layer (μF cm^{-2})
C_{SC}	space charge layer capacitance (μF cm^{-2})
C_{SS}	surface state capacitance (μF cm^{-2})
d_W	space charge layer thickness (cm)
D	diffusion coefficient (cm^2 s^{-1})
E	potential (V)
E_{fb}	flatband potential (V)
e_o	electronic charge (1.602×10^{-19} C)
F	Faraday constant (96 484 C mol^{-1})
f_o	normalisation frequency (Hz)
i	current density (μA cm^{-2})
i_{corr}	corrosion current density (μA cm^{-2})
j	imaginary unit, $j = \sqrt{-1}$
k	Boltzmann constant (1.3807×10^{-23} J K^{-1})
m	mass (g)

M	molar mass (g mol^{-1})
n	exponent of constant phase element
n_{dl}, n_{W}	exponents of CPE_{dl} and CPE_{W}
N	doping density (cm^{-3})
N_A	Avogadro's number (6.022×10^{23} mol^{-1})
N_{SS}	surface state density (cm^{-2})
q	charge density (C cm^{-2})
q_{SS}	charge density associated with surface states (C cm^{-2})
Q	coefficient of CPE (Ω^{-1} s^n cm^{-2}, where n is CPE exponent)
Q_{dl}, Q_W	coefficients of CPE_{dl} and CPE_W (Ω^{-1} s^n cm^{-2})
r	roughness factor
r_{O2}	oxygen molecular diameter (m)
R	universal gas constant ($R = 8.314\ 41$ J K^{-1} mol^{-1})
R_a	adsorption resistance (Ω cm^2)
R_e	solution resistance (Ω cm^2)
R_i	mean integral value of polarisation resistance defined by equation 7.11 (Ω cm^2)
R_n	noise resistance (Ω cm^2)
R_P	polarisation resistance (Ω cm^2)
t	time (s)
T	absolute temperature (K)
V	loss capacitance equation 7.12 (μF cm^{-2})
z	electrons transferred in corrosion reaction
z_i	charge of an ion i
Z^*	complex impedance (Ω cm^2)
Z'	real part of impedance (Ω cm^2)
Z''	imaginary part of impedance (Ω cm^2)
Z_{CPE}	impedance of constant phase element (Ω cm^2)
Z_W	Warburg impedance (Ω cm^2)
$Z(\infty)$	high frequency limit of electrode impedance (Ω cm^2)
β_a, β_c	anodic and cathodic Tafel slope (V dec^{-1})
ΔC	constant, equation 7.4 (Ω^{-1} s^n cm^{-2})
ε_o	dielectric permittivity of vacuum (8.854×10^{-14} F cm^{-1})
ε_r	dielectric constant (relative permittivity)
η	dynamic viscosity (Pa s)
σ	Warburg coefficient (Ω s$^{-0.5}$ cm^2)
$\sigma(I)$	standard deviation of current noise signal (μA)
$\sigma(V)$	standard deviation of potential noise signal (mV)
Φ	phase shift of impedance (deg)
$\phi(sc)$	potential distribution across the space charge layer (V)
ω	angular frequency, $\omega = 2\pi f$ (s^{-1})
ω_o	normalisation angular frequency (s^{-1})

Acknowledgement

The presented study was supported by the grant MSM6046137304 from the Ministry of Education, Czech Republic. The authors thank Dr M. Bartoš for his critical reading of the manuscript.

References

1. A. Turnbull and M. Psaila-Dombrowski, *Corros. Sci.*, 33 (1992), 1925.
2. J. R. Park and D. D. Macdonald, *Corros. Sci.*, 23 (1983), 295.
3. I. Betova, M. Bojinov, A. Englund, G. Fabricius, T. Laitinen, K. Mäkelä, T. Saario and G. Sundholm, *Electrochim. Acta*, 46 (2001), 3627.
4. M. Bojinov, T. Laitinen, K. Mäkelä, T. Saario and P. Sirkiä, in Proc. 4th Int. Symp. on EIS, 393. Rio de Janeiro, Brazil, 1998.
5. T. Dorsch, R. Kilian and E. Wendler-Kalsch, *Mater. Corros.*, 49 (1998), 659.
6. M. Bojinov, T. Buddas, M. Halin, P. Kinnunen, T. Laitinen, K. Mäkelä, T. Saario, P. Sirkiä, K. Tompuri and K. Yliniemi, in Proc. Conf. Chimie, 2002: Water Chemistry in Nuclear Reactor Systems, Soc. Franc. D'Energie Nuc., 142. Avignon, France, 2002.
7. B. Sala, H. Takenouti, S. Chevalier, M. Keddam and A. Gelpi, in Proc. 7th Int. Symp. EMCR, paper 081. Budapest, Hungary, 2000.
8. M. F. Montemor, M. G. S. Ferreira, N. E. Hakiki and M. Da Cunha Belo, *Corros. Sci.*, 42 (2000), 1635.
9. J. Robertson, *Corros. Sci.*, 32 (1991), 443.
10. J. E. Forrest and J. Robertson, *Corros. Sci.*, 32 (1991), 541.
11. D. Lister, in Proc. EUROCORR 2003, European Corrosion Federation Congress, paper 426. Budapest, Hungary, 2003.
12. J. F. Chen and W. F. Bogaerts, *Corros. Sci.*, 37 (1995), 1839.
13. Z. Kerner and T. Pajkossy, *Electrochim. Acta*, 47 (2002), 2055.
14. A. N. Frumkin and V. I. Melik-Gaykazyan, *Dokl. Akad. Nauk*, 5 (1951), 855.
15. M. Růžičková, Ph.D. Thesis, Inst. Chem. Technol., Prague, CR, 2004, 36, in Czech.
16. A. K. Jonscher, *Nature*, 253 (1975), 717.
17. A. K. Jonscher, *J. Mater. Sci.*, 16 (1981), 2037.
18. C. Bataillon and S. Brunet, *Mater. Sci. Forum*, 111–112 (1992), 36.
19. P. Barberis and A. Frichet, *J. Nucl. Mater.*, 273 (1999), 182.
20. J. J. Vermoyal, A. Frichet, L. Dessemond and A. Hammou, *Electrochim. Acta*, 45 (1999), 1039.
21. P. Schmuki and H. Böhni, *J. Electrochem. Soc.*, 139 (1992), 1908.
22. P. Schmuki and H. Böhni, *J. Electrochem. Soc.*, 141 (1994), 362.
23. *CRC Handbook of Chemistry and Physics*, ed. D. R. Lide, 75th edn. CRC Press Inc., Boca Raton, FL, 1994, 12.
24. M. G. S. Ferreira, N. E. Hakiki, G. Goodlet, S. Faty, A. M. P. Simoes and M. Da Cunha Belo, *Electrochim. Acta*, 46 (2001), 3767.
25. D. Wallinder, J. Pan, C. Leygraf and A. Delblanc-Bauer, *Corros. Sci.*, 41 (1999), 275.
26. O. Šifner and R. Mareš, *IAPWS-IF97 Calculations of Thermodynamic, Transport and Other Properties of Plain Water*, CZ NC PWS, Institute of Thermomechanics, Acad. Sci. CR, 2001, in Czech.
27. H. O. Finklea, *Semiconductor Electrodes (Studies in Physical and Theoretical Chemistry)*, vol. 55. Elsevier, Amsterdam, 1988.
28. D. D. Macdonald, *J. Electrochem. Soc.*, 139 (1992), 3434.
29. B. Beverskog, M. Bojinov, A. Englund, P. Kinnunen, T. Laitinen, K. Mäkelä, T. Saario and P. Sirkiä, *Corros. Sci.*, 44 (2002), 1901.
30. Z. Szklarska-Smialowska, *Corros. Sci.*, 44 (2002), 1143.
31. P. Schmuki and H. Böhni, *Electrochim. Acta*, 40 (1995), 775.
32. J. Macák, P. Sajdl, R. Novotný, P. Kučera and Z. Cílová, in Proc. EUROCORR, 2003, European Corrosion Federation Congress, paper 225. Budapest, Hungary, 2003.
33. B. Beverskog and I. Puigdomenech, *Corros. Sci.*, 38 (1996), 2121.
34. B. Beverskog and I. Puigdomenech, *Corros. Sci.*, 39 (1997), 43.
35. K. Rajeshwar, in *Encyclopedia of Electrochemistry*, ed. A. J. Bard and M. Stratmann, vol. 6 *Fundamentals of Semiconductor Electrochemistry and Photoelectrochemistry*, Wiley-VCH 1, 2002, 13.

36. K. Chandrasekaran, R. C. Kainthla and J. O'M. Bockris, *Electrochim. Acta*, 33 (1988), 327.
37. C. Gabrielli and M. Keddam, *Corrosion*, 48 (1992), 794.
38. J. Robertson, *Corros. Sci.*, 29 (1989), 1275.
39. *NACE Corrosion Engineer's Reference Book,* 2nd edn, ed. R. S. Treseder, coed. R. Baboian and C.G. Munger, National Association of Corrosion Engineers, Houston, USA, 1991.
40. H. Göhr, J. Schaller and C. A. Schiller, *Electrochim. Acta*, 38 (1993), 1961.
41. J. R. Macdonald, *Impedance Spectroscopy,* Wiley & Sons, NY, USA, 1987, 90.
42. J. Macák, P. Sajdl, P. Kučera and R. Novotný, in Proc. of Corrosion, 2005, paper 5336. NACE Conference, Houston, USA, 2005.
43. T. Pajkossy and L. Nyikos, *Electrochim. Acta*, 34 (1989), 171.
44. B. Sapoval, *Solid State Ionics*, 23 (1987), 253.
45. A. Einstein, *Ann. Phys.*, 17 (1905), 549.

Monitoring of stress corrosion cracking of sensitised 304H stainless steel by electrochemical methods and acoustic emission

Wenzhong Zhang, Lucia Dunbar and David Tice

Serco Assurance, Walton House, Risley, Warrington, Cheshire, WA3 6GA, UK
wenzhong.zhang@sercoassurance.com; david.tice@serco.com

8.1 Introduction

Stress corrosion cracking (SCC) is one of the most damaging corrosion problems in a variety of industries including the nuclear industry. Unpredicted failures in service may cause safety, economic, environmental and health consequences. In many cases, SCC has a relatively long initiation stage followed by an early stage of short crack growth and coalescence; finally, large cracks appear and rapid crack propagation may lead to a catastrophic failure [1]. Lifetimes of structural components subject to SCC exhibit a large degree of scatter under the same conditions since initiation of SCC is dominated by localised surface condition and localised stress, solution and material conditions, thus presents stochastic characteristics. Development of in-situ monitoring methods, which can detect SCC in its initiation stage and the early stage of short crack growth, can make a considerable contribution to safety management of critical components in plants. Developing experimental methodologies that enable real time monitoring of SCC initiation and early stage propagation is also highly desirable from a mechanistic research point of view.

The thiosulphate/sensitised 304 stainless steel cracking system is interesting for both mechanistic research and practical failure and lifetime evaluation [1–7]. The present research focuses on the SCC initiation and propagation of sensitised 304H stainless steel in dilute tetrathionate solutions at ambient temperature using hybrid electrochemical methods and Acoustic Emission (AE) monitoring techniques.

Useful correlations between Electrochemical Noise (EN) and SCC behaviour have been reported in a number of studies [4–10]. Such correlations could be utilised for detection and evaluation of SCC events. However, automatic real time transient detection as the test unfolded and calculation of associated charge have not been reported.

The research described in this paper employed the Localised Corrosion Monitoring (LCM) analysis software developed by ACM Instruments to detect potential transients and calculate charges while electrochemical measurement is running. The LCM technique records and detects potential transients associated with localised corrosion, including SCC. The advantage of LCM is that it detects and analyses potential transients automatically. Therefore, it has the potential to be developed to a simple method to achieve in-situ real time monitoring of localised corrosion events.

Table 8.1   Chemical composition of 304H (in wt.%)

C	Si	Mn	P	S	Ni	Cr	Mo	Cu	Fe
0.059	0.35	1.10	0.028	0.021	8.55	18.11	0.44	0.49	Bal.

The AE technique provides a unique method of recognising when and where active defects are growing; it has been widely utilised in both laboratories and field applications [10,11]. In the area of SCC study, AE has been used to monitor a large variety of materials including stainless steels, high strength steels, mild steels, as well as aluminium, magnesium, titanium, zirconium and uranium alloys [10–14]. The present research employed two AE sensors with the aim of detecting the timing and location of SCC initiation and propagation.

8.2   Experimental set-up

Commercial-purity 304H stainless steel was used in this study. The chemical composition of the material is shown in Table 8.1.

The material contains a relatively high level of carbon (0.059 wt.%) and is susceptible to thermal sensitisation. After solution heat treatment (1050°C/0.5 h, nitrogen cooling), the material was sensitised at 600°C for 100 h. The specimens were then machined out of 20 mm diameter sensitised 304H rods. The gauge section of the specimen includes two 20×20 mm flat surfaces and two 20×6.38 mm curved side surfaces, the total gauge section surface area being approximately 1055 mm^2. There are four sharp edges between the flat and side surfaces. The specimens were coated with Lacomit lacquer to mask the surface of the round rod sections and the shoulders between the round rod and the gauge section. This ensured that only the gauge section was exposed to test solution.

Figure 8.1 shows the experimental set-up. The load machine is an INSTRON 50 kN materials test machine. For LCM and EN monitoring, a Gill 8 from ACM Instruments was used. The specimen was used as working electrode 1 and connected to the WE1 input. The WE2 input was connected to a 6.3 mm diameter sensitised 304H rod acting as working electrode 2. The surface area exposed to test solution was approximately the same for WE1 and WE2. The reference electrode (RE) was an unsensitised 304 wire with PTFE sheath. The counter electrode (CE) was a Pt flag which was optional for LCM. The sampling rate for potential and current recording used throughout the tests was 5 Hz.

A PCI-2-based AE system from Physical Acoustics Corporation with two R15 AE sensors was used for AE monitoring. The two AE sensors were pressed onto the two ends of the specimen with two PTFE sheaths and were housed in two high strength steel specimen grip tubes. Sensor outputs were amplified to 40 dB by pre-amplifiers.

The specimen was normally first loaded to about 2 MPa stress to engage the load fixture, then test solution was added to the PTFE container. Shortly after, the stress was increased to either 180 MPa or 90 MPa and kept at constant load. The test solution was 0.5 wt.% potassium tetrathionate (pH at about 2.96 adjusted with H_2SO_4). All of the tests were run at room temperature.

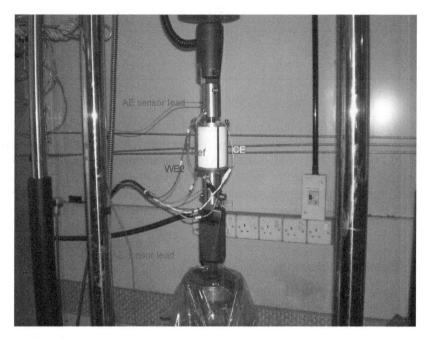

8.1 Experimental set-up

8.3   Results and analysis for electrochemical noise

8.3.1   Potential and current transient response to SCC initiation and propagation

A typical result of a sensitised 304H specimen is shown in Fig. 8.2. The specimen (designated as R614) was tested at 23 kN constant load, equivalent to a 180 MPa stress for the cross-section of the specimen in the central exposed section.

The upper diagram in Fig. 8.2 is the recording of current (mA) versus time. The bottom diagram of Fig. 8.2 is the recording of potential (mV) versus time (s). It should be noted that the LCM software allows the detection and charge calculation for each transient to be done while the test is running as well as after the test.

From Fig. 8.2, it can be seen that, firstly, a number of relatively small potential transients were detected. Then, a relatively large potential drop (about 200 mV) and current jump (about 15 µA) were observed at around 18 000 s followed by seven other large potential drops and corresponding current jumps. Between these eight relatively large transients, several smaller transients were also detected. At roughly 67 000 s, the potential started to decrease steadily whilst the current steadily increased. At about 84 000 s, the specimen broke, in correspondence to a sharp potential drop and current jump which happened over a very small timescale.

Figures 8.3 to 8.6 show enlarged views of interesting sections of the EN/LCM record in Fig. 8.2. Figure 8.3 shows that, when the specimen was loaded to 23 kN, a small potential drop (about 10 mV) and a small current jump (about 0.25 µA) were detected. From Figs 8.2 to 8.6, it can also be seen that potential drops and current jumps were always detected simultaneously.

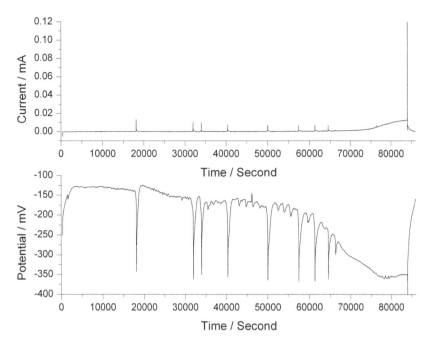

8.2 Potential and current transient recorded for specimen R614

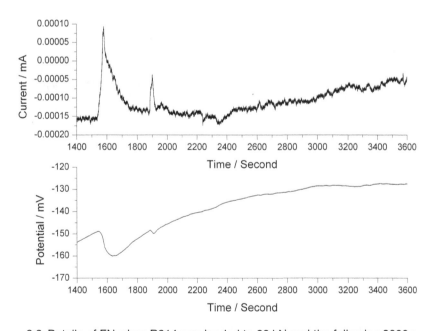

8.3 Details of EN when R614 was loaded to 23 kN and the following 3000 s

8.3.2   Interpretation of R614 results

The potential and current transients are all linked with the electrochemical reactions that take place on the material surface and the aspect of the EN records in Figs 8.2 to 8.6 can be explained as follows.

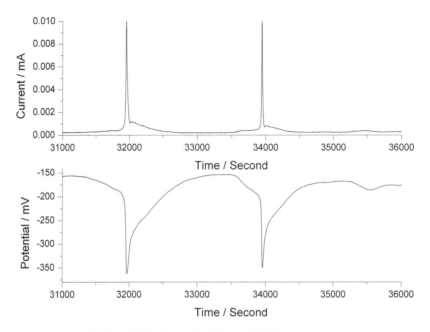

8.4 Details of EN for R614 from 31 000 to 36 000 s

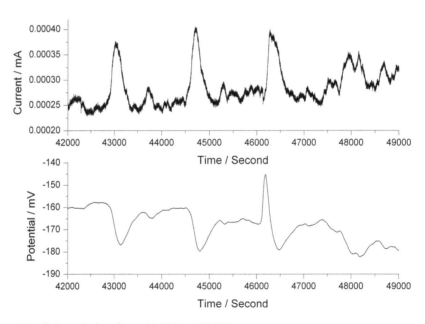

8.5 Enlarged view from 42 000 to 49 000 s

When the electron consumption speed of cathodic reactions is in balance with the speed of electron generation by anodic reactions, as is the case for uniform corrosion, the corrosion potential remains relatively stable, or goes up slowly and smoothly because of passivation. However, when localised corrosion events happen, as in SCC,

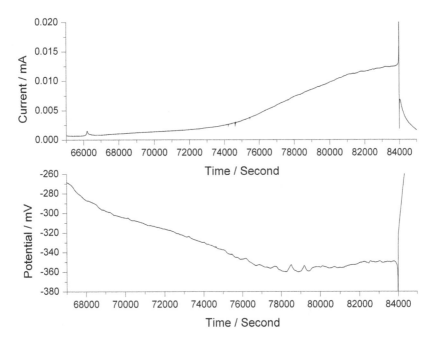

8.6 Details of longest transient until R614 breaks at 84 000 s

the excess electrons produced by a localised corrosion event will disturb the balance, producing a sudden burst of potential transient.

In general, excess electrons produced by localised corrosion (anodic reactions) are either directly consumed by the cathodic reactions or accumulated temporarily in the double-layer capacitance. When the electron consumption rate of cathodic reactions is slightly behind the rate of electron generation by the anodic reactions, a slow potential drop occurs to balance the two rates.

The rate of electron production by localised corrosion can be relatively fast and, in this case, the cathodic reactions cannot consume the excess electrons immediately, so the excess electrons charge up the double-layer capacitance rapidly and cause a steeper potential drop. In the case of SCC, if a large fresh metal surface is suddenly exposed to solution, the excess electrons produced by anodic reactions are sufficient to cause an abrupt potential drop. From Fig. 8.4, it can be seen clearly that the large potential transients always start with a slow potential drop followed by a faster potential drop, and some of them end with an abrupt potential drop, which may indicate that an SCC event involved a relatively large area.

This suggests that the SCC event might start with a relatively slow anodic dissolution reaction, followed by the anodic reaction speeding up, and, in some cases, a relatively large cracking event occurs which causes a large and abrupt potential drop. A potential drop of this magnitude may be due either to SCC crack coalescence or to propagation of a relatively large crack whereby a large area in front of the crack tip becomes weak enough for the stress to cleave it apart suddenly. Depending on the corrosion or re-passivation processes at the freshly cleaved surface and on the mass transportation effects in the solution local to the crack, the anodic reactions after an SCC event would normally dampen down. The excess electrons stored in the double-layer capacitance are then gradually consumed by cathodic reactions which in turn would cause the potential to recover. Conversely, when there is continued supply of excess electrons produced by localised corrosion events or fresh cracking surface areas, the potential will drop steadily as is suggested in Fig. 8.6. It is likely that, at this stage, crack propagation was rapidly producing fresh crack surfaces on a continuous basis with excess electrons to charge up the double-layer, so in this case the cathodic reactions could not keep pace with the anodic reactions continuously, thus the potential dropped steadily and current increased continuously.

Considering that 180 MPa is less than the 0.2% yield stress at room temperature of 205 MPa for the material of interest, one could suggest that the EN records are indicative either of stress intensification where localised stress was above yielding or of brittle cracking without apparent yielding. The sharp corner edges between flat and side curved surfaces and scratch-like defects on side surfaces within the area exposed to the solution may be amongst the possible causes of stress intensification.

8.3.3   Estimate of charge transferred during localised corrosion events for R614

Since the material and surface area of WE1 and WE2 are the same and the solution conductivity is high, it would be expected that half of the excess electrons produced by localised corrosion events on WE1 will move to WE2 through the Zero Resistance Ammeter (ZRA) which measures the current transients. Electrode WE2 was not subjected to stress and had a better polished surface, thus it can be expected to have had a much smaller tendency for localised corrosion. The charge transferred during the transients could be directly linked with the amount of consumed metal in the specimen during the detected localised corrosion events. Figure 8.7 shows the number of detected current transients and their calculated charges.

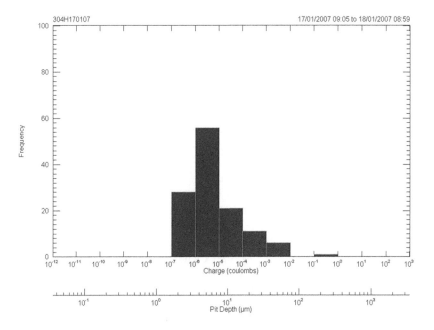

8.7 Number of current transients and their associated charges for R614

8.8 Photograph of specimen R 614 after SCC test

The notional 'pit depth' in μm can be calculated from the corresponding charge transferred in an event based on a simple pit corrosion model defined in the LCM software. For example, a charge between 0.001 and 0.01 Coulombs corresponds to a pit depth of roughly 30–80 μm, e.g. an average pit depth of approximately 60 μm, which means such detected charge is expected to produce on average a pit of $60 \times 60 \times 60$ μm^3, or expressed in another way, a volume of dissolved metal that can produce a current transient with such a charge is 2.16×10^{-4} mm^3. Obviously, it is virtually impossible to produce such a large pit in one single corrosion event. It is likely, therefore, that events involving this amount of charge are SCC events.

Adding together all of the detected transients and their charges, an estimated 0.0172 mm^3 volume of metal was consumed. On the further assumption that an SCC crack of 6.38 mm in length, that is spanning the full width of the curved side of the specimen, and 80 μm in width, corresponding to roughly the average grain size, the calculated penetration depth of the SCC crack will be approximately 33.7 mm. Figure 8.8 shows a image of the broken specimen R614.

There is one main through crack and over 20 much smaller cracks, which are visible to the naked eye. If the 20 smaller cracks are counted roughly as half a through crack, the total penetration depth of these visible cracks amounts to about one and half times the width of the flat surface, e.g. $1.5 \times 20 = 30$ mm, which is close to the calculation given above based on charge estimation.

8.3.4   Additional tests interrupted after first large transient

Since a large transient corresponded to SCC with a volume of about $60 \times 60 \times 60$ μm^3, which would be a practical detectable crack, the following test (specimen R628) was stopped as soon as the first large potential drop and associated current transients were detected.

One of the curved side surfaces of specimen R628 was photographed before the test. The image is shown in Fig. 8.9. Some light scratches can be seen clearly on the side surface. Figure 8.10 shows an enlarged view of a few of the scratches.

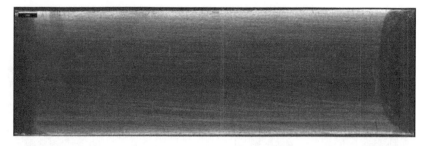

8.9 Scratch-like defects on one of the curved sides of specimen R628 before test

8.10 Enlarged view of defects on specimen curved side

Figure 8.11 shows the recorded potential and current transients for this specimen with Fig. 8.12 showing sections of the records in greater detail. From Figs 8.11 and 8.12, it can be seen that, when loading to 23 kN at around 1700 s, a small potential drop (about 20 mV) and a small current jump (about 0.28 µA) were detected. A relatively large potential drop (about 125 mV) and current jump (about 4.5 µA) were detected at around 4000 s.

The large transient in Fig. 8.11 indicated that the potential started to drop slowly at around 3300 s, and that after a further 500 s, the potential drop became faster. After another 250 s, the potential dropped abruptly. This was interpreted as indicating a fast SCC event at around 4150 s which was probably the result of the gradual build up of a slow process which had started at about the 3300 s.

An interesting finding from Fig. 8.12 is the relatively small transient (a potential drop of about 0.3 mV and a current jump of about 0.05 µA) that was observed before the load was applied. This is likely to have been caused by inter granular attack (IGA) alone, with no contribution of any SCC mechanism. When the following

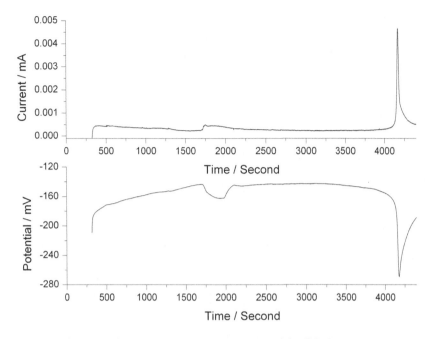

8.11 EN records (potential and current transients) for R628

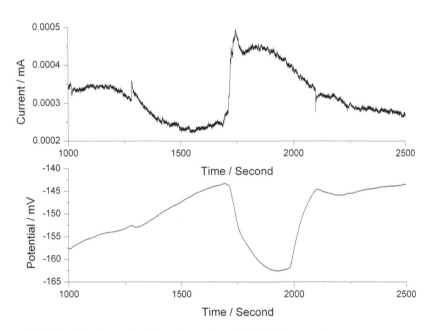

8.12 Details of selected transients from R628 ECN record

transient, caused by applying the load, was still in the recovery stage, a relatively small transient was detected at around 2100 s. This transient is noteworthy because it is quite different from other transients: it shows an abrupt current drop rather than the usual jump, which suggests that it may have been caused by an IGA event on the second working electrode WE2, whereby this electrode supplied electrons rather than taking electrons up. Although this transient is identified as a negative current transient, the LCM software counts this as a normal current transient because LCM software only detects and counts potential drops, irrespective of current direction. This example shows how important it is to choose the second working electrode carefully ensuring that local corrosion on it is minimised so as not to interfere with monitoring of the main WE1 electrode (the specimen under study).

8.3.5   Analysis of charge associated with transients for R628

Figure 8.13 shows the number of detected transients and their calculated charges from the EN record of specimen R628.

The 0.0001 to 0.001 Coulomb charge transients with a frequency of 2 correspond to the two major transients detected; one is caused when applying load, and the other is likely to have been caused by SCC. Although the transient caused by applying the load is small in magnitude, because of its long duration, it has been associated with a total charge comparable with the much more dramatic potential transient observed later. The other three very small transients with charges less than 0.00001 Coulombs are most likely caused by IGA, with one such event occurring on WE2 as discussed before.

For the two major transients, the 0.0001 to 0.001 Coulombs charge corresponds to an average penetration depth of approximately 25 μm, that is an estimated $25 \times 25 \times 25 \ \mu m^3 = 1.56 \times 10^{-5} \ mm^3$ of metal was consumed to produce a current transient with such a charge.

The test was stopped when the first large correlated potential and current transient was detected and the specimen surface was examined under the optical microscope:

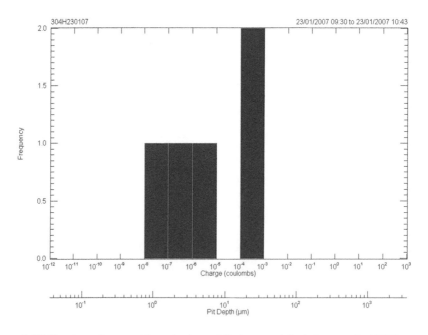

8.13 Number of current transients and their associated charges for R628

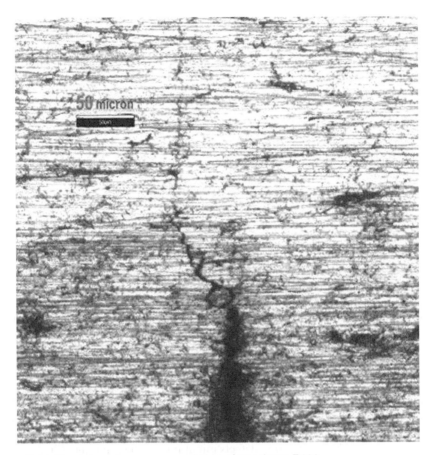

8.14 Crack found on one curved side of specimen R628

only one significant crack was found. Figure 8.14 shows the crack found on the specimen surface, which was on one of the curved sides of the specimen.

It can be seen clearly from Fig. 8.14 above that a stress corrosion crack initiated from the sharp tip of a pre-existing defect. Measured from the tip of the defect, the crack is approximately 80 μm in length and 1–3 μm in width. Ahead of this crack, there are a few much smaller cracks with width less than 1 μm and total length of about 80 μm.

If we assume that the crack initiated when load was applied and it had undergone an early stage growth or coalescence when the first relatively large transient was detected, then the volume of the cracks would be approximately $2 \times 25 \times 25 \ \mu m^3$ or $3.13 \times 10^{-6} \ mm^3$. If we use 160 μm as the crack length and 1 μm as the crack width, the estimated penetration depth of the cracking is approximately 98 μm, which is about one or two grains in depth. This compares well with other published results [5,6], which reported cracks of length between 60 and 300 μm for similar material in a similar test solution. This corresponds to a crack of the size of one or more grains for a single crack event, which is considerably larger than would be expected for cracks resulting from slip-dissolution mechanisms, estimated to be of the order of 0.001 to 0.1 μm [5]. This crack size is consistent with the size of a crack one could expect to be produced because of a hydrogen-induced brittle fracture event occurring across the crack front, e.g. due to hydrogen embrittlement [5,6].

8.3.6   Mechanisms at play in the observed SCC cracks

Considering in detail the chemical nature of the solution used in the tests, it is noted that tetrathionates and thiosulphate are coupled by the following redox reaction [15]

$$S_4O_6^{2-} + 2e^- = 2S_2O_3^{2-} \tag{8.1}$$

The nature of the thiosulphate anion is such that it supplies sulphur near an incipient crack, where it is adsorbed and accumulates, via chemical disproportionation or electrochemical reaction [5,6,15]

$$S_2O_3^{2-} + H^+ = S + HSO_3^{2-} \tag{8.2}$$

or

$$S_2O_3^{2-} + 6H^+ + 4e^- = 2S + 3H_2O \tag{8.3}$$

The detailed analysis of the potential transients above, in combination with what was reported in a number of published papers [5,6], points to the following mechanism. First, the SCC starts with slow anodic dissolution at the chromium-depleted grain boundaries. Subsequently, elemental sulphur is adsorbed on the surface around the crack tip and catalyses the entry of hydrogen atoms produced by the hydrogen reduction reaction into the steel matrix ahead of the crack tip where it accumulates gradually over several hundred or thousand seconds. This eventually causes hydrogen-induced brittle fracture in the steel. Strain-induced martensite ahead of the crack tip may contribute to the enhanced hydrogen atom diffusivity and permeability [16]. Hydrogen may also locally enhance the plasticity of the matrix ahead of the crack [17]. Hydrogen may be preferentially trapped at the carbide/matrix interface thus reducing the strength of the interface [18]. Complex stress/strain conditions caused by pre-existing defects might favour the trapping of hydrogen. All of these factors may contribute to the hydrogen-induced brittle fracture in the steel.

8.3.7   Interrupted test completion for R628

Following the microscopic examination, specimen R628 was put back on the test rig. Figure 8.15 shows the full EN record of transients detected for this specimen. After resuming the test and reloading to 11.5 kN, which is only half the load applied before the test was interrupted, a significant potential drop and a current increase were detected. This again indicated a possible stress intensification effect. The specimen was maintained at a load of 11.5 kN (or 90 MPa stress) for a period of time when a series of potential drops and current jumps of varying sizes were detected. It should be noted that the LCM software counted the stoppage time from about 4400 to about 21 000 s together with the first detected large transient as a single transient with a long duration. This is an artefact of the LCM software which will be corrected in charge calculation. When smooth and continued potential drop and current increase started to appear, the experiment was stopped again and the specimen was again examined under the microscope.

Two large cracks and nine small cracks were found on the side surfaces, and front and back flat surfaces. The large cracks were imaged using an Alicona 3D microscope, as shown in Figs 8.16 and 8.17. Figure 8.16 shows clearly the presence of several of the cracks which initiated from the pre-existing defects, but not all of them

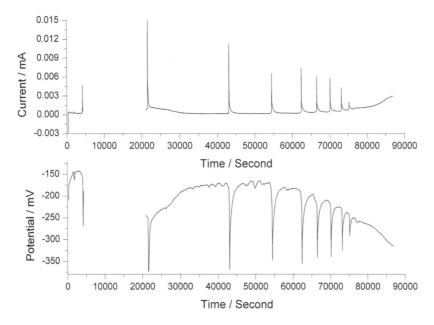

8.15 Full EN record of transients detected for specimen 628

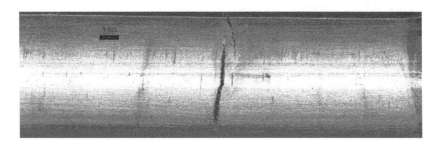

8.16 Large cracks on the curved side of specimen 628

initiated SCC. The widest crack was the one found when the test was interrupted at about 4400 s.

8.3.8 Analysis of charge sizes for specimen R628

Figure 8.18 shows the number of detected current transients with their associated calculated charges.

As already mentioned above, the LCM software detected a transient of 0.1 to 1 Coulombs combining the stoppage time, which is one of the transients counted in this charge range. This should have been measured as a transient with charge of between 0.0001 and 0.001 Coulombs. After correction and adding together the detected transients and their charges, it was estimated that a 0.0174 mm³ volume of metal was consumed in this transient. As before, assuming that the SCC crack was 6.38 mm in length, that is the width of the curved side surface, and 80 μm in width, which is about an average grain size, the calculated penetration depth of the SCC

8.17 Large cracks on flat surfaces, seen from either side of specimen R628

crack would be approximately 34 mm. The two large cracks and nine small cracks observed were found to have roughly such a depth.

From Fig. 8.18, it can be seen that small transients, associated with charges less than 10^{-8} Coulombs and corresponding metal dissolution of about 0.125 μm^3, could be detected. For transients this small, it would be difficult to claim that it is possible to distinguish SCC from other localised corrosion events due to other mechanisms.

When the charge estimated exceeds 10^{-5} Coulombs in one single event, it is quite likely that a SCC event has taken place, because this can be associated with metal dissolution of over 1000 μm^3 in a single event.

It is noted that in such a corrosive solution as used in our tests, it takes several hundred seconds for hydrogen atoms to accumulate ahead of the crack tip in sufficient quantity to cause hydrogen-induced cracking. It can be envisaged that, in real conditions with much less corrosive environments, it would take much longer to accumulate enough hydrogen for this mechanism to operate. There should also be sites such as the carbide/matrix interface or complex stress/strain areas available for preferential trapping of hydrogen to cause hydrogen-induced brittle fracture, otherwise hydrogen will not accumulate but will diffuse away rapidly.

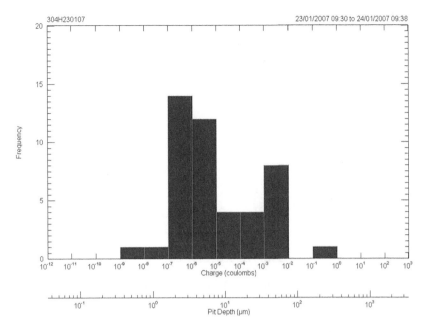

8.18 Number of detected current transients and their calculated charges

8.4 Results and analysis for acoustic emission

8.4.1 AE results for R614

Typical AE results for a sensitised 304H specimen are shown in Figs 8.19–8.21. The specimen is R614 and, as previously mentioned, it was tested in 0.5 wt.% potassium tetrathionate solution, at pH 2.96 and room temperature, and under a 23 kN constant load (180 MPa stress).

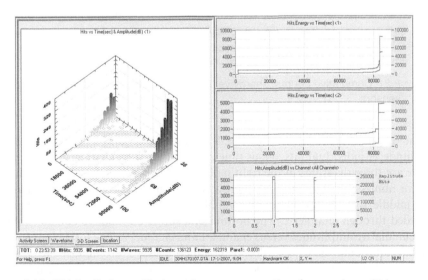

8.19 AE hits, their amplitude and energy versus time for specimen 614

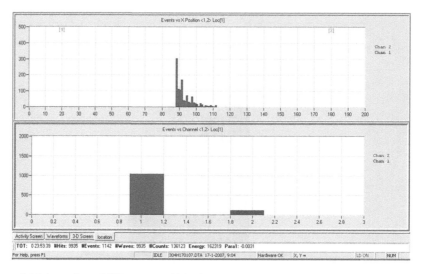

8.20 Location of AE events within the area exposed to the solution for R614

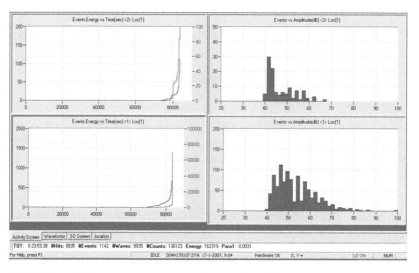

8.21 Cumulative counts, energy vs time and amplitude distributions for both sensors

The diagram in the left half of Fig. 8.19 shows the number of AE hits and the amplitude of individual hits versus time recorded through channel 1 (AE sensor 1). There are three diagrams in the right half of Fig. 8.19. The top one shows the cumulative number of AE hits and energy versus time recorded through channel 1. The middle one shows cumulative number of AE hits and energy versus time recorded through channel 2. The bottom one shows cumulative counts of AE hits and cumulative amplitudes versus channel 1 and channel 2. The total numbers of hits and the total amplitudes recorded by channels 1 and 2 are almost the same, and this was taken to indicate that both AE sensors worked at similar sensitivity such that most of the AE signals could be detected by both sensors.

From Fig. 8.19 it can be seen that, when loading to 23 kN and when the specimen was approaching fracture, bursts of AE hits were detected but in between, only a few scattered AE hits were recorded.

8.4.2   Analysis of event location for R614

Figure 8.20 shows the location of AE events recorded. Only AE hits which can be picked up by both sensors within a specified time delay are counted as 'events'. This time delay is calculated based on the travel time required for sound to reach the sensors from the exposed area of the specimen along the specimen length so that noise signals generated by equipment fixtures and the surrounding laboratory could be easily filtered out.

The top diagram in Fig. 8.20 shows the relative position of sensor 1 (20 mm) and sensor 2 (180 mm) with respect to the exposed area of the specimen (from about 90 to 110 mm). In the bottom diagram of Fig. 8.20, the bars on the left show the number of events detected by sensor 1 first, whilst the bars on the right show the number of events detected by sensor 2 first.

From Fig. 8.20, it can be seen clearly that the majority of AE events detected were at the end of the exposed area nearest to sensor 1. The microscopy image of the specimen after test has been shown in Fig. 8.8. Considering that sensor 1 was fixed to the left end of the specimen shown in Fig. 8.8, the location of AE events indicated by Fig. 8.20 is in good agreement with the location of the cracks observed on specimen R614.

8.4.3   Interpretation of AE results for R614

Our test results indicate that AE seems sensitive to rapid crack propagation but not to SCC initiation and early stage propagation for this kind of material–environment–load combination. Figure 8.21 shows the accumulated number of AE events and their energies versus time (the left half of the figure) as well as the AE event amplitude distributions (the right half of the figure) for both sensors.

It can be seen from Fig. 8.21 that AE events started being picked up from around 70 000 s and the specimen fractured at around 84 000 s. In comparison with the LCM result for the same specimen shown in Fig. 8.2, it can be seen that a continuous stream of AE events start to be detected when the potential starts to drop steadily whilst the current steadily increases. This suggests that a stream of AE events can only be detected when a crack has reached the stage of continuous propagation with large scale plastic deformation.

It is postulated that the AE technique is sensitive to SCC events when these involve relatively large plastic deformation ahead of the SCC crack tip and not to SCC events involving extremely small area or virtually no plastic deformation. At the later stage, SCC crack advance can be expected to have reduced the cross-sectional area at the level of the exposed section of the specimen to a point where the stress across the ligament has exceeded yielding stress values. This is where AE events start to be detected in a continuous stream. The AE event amplitude distribution also gives a clue in this direction since peak amplitudes below 48 dB have been linked with AE signals originating from plastic deformation [13].

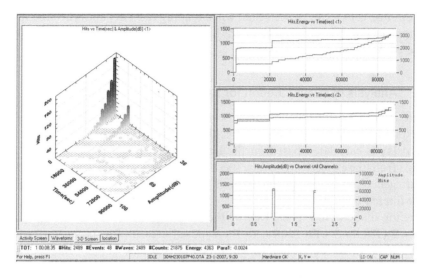

8.22 AE hits, their amplitude and energy versus time for specimen R628

8.4.4   AE results for specimen R628

Figures 8.22 to 8.24 show the AE results for the interrupted test of specimen R628.

From Figs 8.22 and 8.23, it can be seen that bursts of AE hits and a few AE events were detected during the increasing of load to 23 kN. During the period at constant load of 23 kN, when there was a first large potential drop and corresponding current jump, as detected using LCM and EN at around 4000 s (Fig. 8.11), there were no corresponding AE hits or events at this time. When the test was resumed, loading the specimen to 11.5 kN, bursts of AE hits were detected. In a period of constant load at 11.5 kN, when a series of potential drops and current jumps of varying sizes were detected (Fig. 8.15), only scattered AE hits could be detected with no AE events at all.

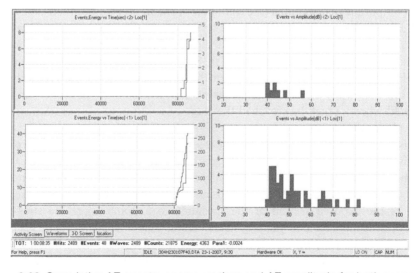

8.23 Cumulative AE events, energy vs time and AE amplitude for both sensors

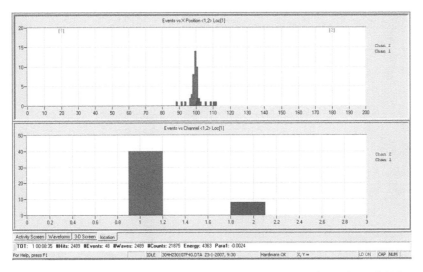

8.24 Location of AE events within the exposed section of specimen R628

Comparing Fig. 8.15 with Fig. 8.23, it can be seen that, when a continuous potential drop and current increase were detected by LCM and EN at about 80 000 s, the accumulated number of AE events began to increase steadily. Figure 8.24 shows the location of AE events for R628 and from this figure, it can be seen clearly that the majority of AE events were detected in a location near the middle of the specimen.

When the experiment was stopped following detection of these large signals, two large cracks were found near the middle of the specimen as shown already in Figs 8.16 and 8.17. Again the location of AE events is in good agreement with the observed location of specimen fracture.

Although there were quite a number of AE hits recorded on both AE sensors whilst the load was kept at 11.5 kN, most of them were not counted as AE events. To see whether increasing the assumed specimen gauge length for accepting more AE hits as events could give useful insight into the results of the tests, the filter thresholds were adjusted so that acoustic emissions anywhere within the specimen would count as events. It was found that many of the AE signals were produced in locations corresponding to the thread connection of the specimen in the cell grips, as can be seen in Fig. 8.25.

8.4.5   Analysis of acoustic emission results for R628

The finding in Fig. 8.25 can be explained considering that acoustic emissions are mainly stress waves produced by sudden movement in stressed materials. In our tests, the applied stress was well below the yielding stress, so we can assume that, at the early stage of SCC initiation and propagation, the specimen was experiencing an elastic stress regime. Although extremely small areas ahead of the crack tips might have experienced intensified stress reaching values above yielding, these could not produce strong enough AE signals above the detection threshold. When SCC cracking had reached a sizeable length, the loading fixture would have to adjust the tension to maintain the load constant, and at this stage the grips moved to adjust their position giving rise to the detected AE hits.

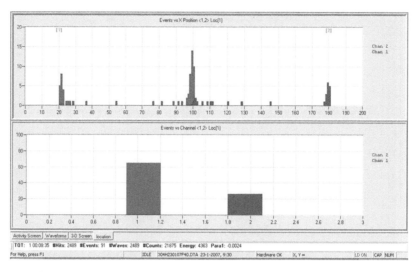

8.25 Location of AE events over the whole length of the specimen

The lack of detection of an AE event in the gauge section of the specimen during the slow SCC propagation stage suggests that there was no large scale plastic deformation during this stage. The plasticity of a small area ahead of the crack tip may be enhanced by hydrogen atoms; stress intensification may cause an extremely localised area of plastic deformation ahead of the crack tip; but AE signals from them were below the detectable threshold. It has been reported that the AE technique is sensitive to a variety of localised phenomena such as: hydrogen evolution on the specimen surface, cracking of corrosion products and grain detachment [8,11]. We can speculate that hydrogen evolution, cracking of corrosion products and grain detachment did not occur during the SCC initiation and early stage propagation in our tests. This is considered likely because sulphur is an element which catalyses hydrogen atom production but hinders the evolution of hydrogen gas; the acidity of the solution would promote metal dissolution rather than producing a thick layer corrosion product.

Conversely, when SCC crack propagation had reached a stage where the stress across the ligament had exceeded yielding, it can be expected that relatively large scale plastic deformation and rapid crack propagation started to occur, giving rise to the continuous stream of AE events which was observed within a period before specimen fracture.

8.5   Conclusions

The findings of this work on sensitised 304H stainless steel in dilute tetrathionate solutions may be summarised as follows:

1. EN measurements: Correlated potential and current transients which are linked with localised corrosion events can be detected using LCM and EN methods. By examining these records, it is possible to separate SCC events from other types of localised corrosion events, as SCC is characterised by relatively large scale drops in potential and simultaneous jumps in current. The potential drop and current jump correlated with SCC events take a relatively long time to recover

and in some cases the potential does not recover to the original values, depending on the corrosion or repassivation processes at the newly formed crack faces, on mass transportation phenomena in the local environment close to the crack and on the speed of cathodic reactions which consume the excess electrons produced by the anodic reactions of the SCC events.

2. Detection of distinct SCC crack evolution stages: LCM and EN methods are highly suitable for the detection of SCC crack initiation and the early stage of propagation. The size of the crack can be estimated based on the estimation of the charge involved in the detected transients. Calculation assuming the spanning of the curved side surface of the specimen as crack length and the average grain size as the crack width yields an estimated penetration depth in relatively good agreement with the crack penetration depth observed. The 'wriggling' nature of the SCC paths was taken into account by assuming the average grain size as the SCC width.

3. The unique shape of large potential transients: The shape of the LCM signal at the start of a dramatic potential drop suggests that this kind of SCC event may start with a relatively slow anodic dissolution reaction, then the anodic dissolution speeds up, and in some cases, a relatively large cracking event occurs which causes an abrupt potential drop.

4. Filtering of AE signals and events location: Filtering of AE events from AE hits, through imposing that only signals reaching both sensors within a given window of time related to specimen gauge size and signal flight times, can effectively eliminate AE signals generated by the load fixtures and the surrounding laboratory. The location of the AE events determined by the two sensors is in good agreement with the location where the final cracks have been observed in the specimen.

5. Specific suitability of AE technique: The AE technique is sensitive to rapid cracking propagation but does not appear to be sensitive to SCC initiation and early stage propagation for the kind of material–environment–load combination used in this study. It is postulated that the AE technique is sensitive to SCC propagation involving relatively large areas of plastic deformation and not to SCC with extremely small areas or virtually no plastic deformation.

6. Mechanism of sensitised 304 SCC in acidic conditions: A mechanism for SCC initiation and early propagation has been proposed, based on the analysis of our experimental results and information from the literature. This postulates that SCC starts with a slow anodic dissolution at the chromium-depleted grain boundaries. Subsequently, elemental sulphur is adsorbed on the surface around the crack tip and catalyses the entry of hydrogen atoms produced by the hydrogen reduction reaction into the steel matrix ahead of the crack tip, where it accumulates gradually over a relatively long time and preferentially at the carbide/matrix interface. This eventually causes hydrogen-induced brittle fracture alongside grain boundaries.

7. Simultaneous use of diverse techniques: Simultaneous use of LCM, EN and AE has proved to be a powerful tool for the detection of SCC initiation and propagation. In addition, it also offers potential for supporting the interpretation of the phenomena which give rise to these records.

Acknowledgements

The authors gratefully acknowledge John Stairmand and Norman Platts for helpful discussions. The authors thank Bernard MacGrath, Will Daniels, Russell Smith, Paul Ogden and Steve Jacques for their help with the INSTRON or AE equipment.

References

1. Y. Watanabe and T. Kondo, *Corrosion,* 56 (2000), 1250.
2. H. S. Isaacs, B. Vyas and M. W. Kendig, *Corrosion,* 38 (1982), 130.
3. R. C. Newman, K. Sieradzki and H. S. Isaacs, *Met. Trans. A,* 13A (1982), 2015.
4. J. K. Lee and Z. Szklarska-Smialowska, *Corrosion,* 44 (1988), 560.
5. D. B. Wells, J. Stewart, R. Davidson, P. M. Scott, D. E. Williams, *Corros. Sci.,* 33 (1992), 39.
6. M. Gomez-Duran and D. D. Macdonald, *Corros. Sci.,* 48 (2006), 1608.
7. T. Haruna, T. Shibata and R. Toyota, *Corros. Sci.,* 39 (1997), 1873.
8. R. A. Cottis and C. A. Lotto, *Corrosion,* 46 (1990), 12.
9. K. Yamakawa and H. Inouc, *Corros. Sci.,* 31 (1990), 503.
10. A. Yonezu, H. Cho and M. Takemoto, *Meas. Sci. Technol.,* 17 (2006), 2447.
11. S. Yuyama, 'Fundamental aspects of acoustic emission applications to the problems caused by corrosion', in *Corrosion Monitoring in Industrial Plants Using Nondestructive Testing and Electrochemical Methods,* 43–74. ASTM STP 908. ASTM, Philadelphia, PA, 1986.
12. R. H. Jones and M. A. Friesel, *Corrosion,* 48(9) (1992), 751.
13. K. Y. Sung, I. S. Kim and Y. K. Yoon, *Scripta Mater.,* 37 (1997), 1255.
14. F. Ferrer, E. Schille, D. Verardo and J. Goudiakas, *J. Mater. Sci.,* 37 (2002), 2707.
15. S. Ritter, T. Dorsch and R. Kilian, *Mater. Corros.,* 55 (2004), 781.
16. G. Han, J. He, S. Fukuyama and K. Yokogawa, *Acta Mater.,* 46 (1998), 4559.
17. T. M. Angeliu, D. J. Paraventi and G. S. Was, *Corrosion,* 51 (1995), 837.
18. G. A. Young and J. R. Scully, *Scripta Mater.,* 36 (1997), 713.

9

Corrosion monitoring of carbon steel in pasty clayey mixture as a function of temperature

Christian Bataillon, Frantz A. Martin and Marc Roy

CEA, DEN, SCCME/LECA, 91191 Gif-sur-Yvette, France
frantz.martin@cea.fr

9.1   Introduction

The French Agency for Nuclear Waste Management (ANDRA) has proposed several concepts to dispose of nuclear wastes inside stable geological layers. For one of these concepts, nuclear waste could be put in a canister in a carbon steel pipe. This pipe would be set in a hole excavated in the host rock. The space between the walls of the hole and the pipe could be filled with engineered sintered clay.

To prevent release of radionuclides, several barriers would be assembled. Stainless steel canisters, carbon steel overpacks, carbon steel pipes and sintered clay are candidate barriers. As carbon steel will corrode in groundwater, the overpack and the pipe could prevent radionuclide release only for a limited time. As clay can trap some radionuclides inside its stacked structure, the sintered clay could act as a barrier after canister–overpack–pipe corrosion breakdown. Since clay exhibits reasonable trapping features for temperatures lower than 80°C, the canister–overpack–pipe system has to act as a barrier until the temperature around the canister decreases sufficiently. For reprocessed waste and spent fuel, the corresponding time needed is at least 1000 years.

This arrangement assumes that the corrosion rate of carbon steel over long periods of time could be estimated and also that the corrosion products should not destroy the trapping features of the sintered clay barrier.

The purpose of this chapter is to give some qualitative features for the corrosion process of carbon steel under conditions similar to those assumed for long-term geological disposal. Qualitative aspects for corrosion monitoring and formation of corrosion products are presented here. In the second part, impedance modelling will be reported and related to quantitative corrosion monitoring.

This work was initiated, followed and supported by ANDRA in the framework of its project on deep geological disposal of nuclear wastes.

9.2   Experimental procedures

9.2.1   Materials and solutions

The working electrodes were made of 1050 carbon steel. The metallurgical structure consists of two phases: ferrite and pearlite grains. Pearlite grains appear striated. It is a lamellar structure consisting of stacked ferrite and Fe_3C thin planes.

The solutions were mixtures at 50% in weight made of 'FoCa' clay and synthetic groundwater. The mineralogical composition of the 'FoCa' clay is given in Table 9.1. This swelling clay is a natural raw material from Normandy, France. Two types of

Table 9.1   Mineralogical composition of 'FoCa' clay

Bentonite	$(Si_{3.6}Al_{0.4}) (Al_{1.75}Mg_{0.15}Fe^{+III}_{0.15})O_{10}(OH)_2 Ca_{0.2}$	80%
Kaolinite	$Si_2Al_2O_5(OH)_4$	4%
Quartz	SiO_2	6%
Calcite	$CaCO_3$	1.4%
Goethite	$FeO(OH)$	6%
Hematite	Fe_2O_3	0.25%
Gypsum	$CaSO_4 \cdot 2H_2O$	0.4%
Water	H_2O	$\approx 9\%$[a]

[a]Depending on the atmospheric hygrometry.

solution were used. One corresponds to a granitic groundwater, the other to a clayey groundwater. The chemical compositions of these solutions are given in Table 9.2.

9.2.2   Reference electrode

Mixtures of clay and groundwater used in this study are pasty. They contain colloids and solid grains of sub micrometric to millimetric size. Colloids and small grains can block the sintered material ending the reference electrode, leading first to the development of an unknown junction potential and second to the breakdown of the liquid junction between the mixture and the internal part of the reference electrode. Consequently, the use of standard reference electrodes is practically impossible for long-term experiments in this pasty mixture.

In a potentiostatic set-up, the reference electrode is used as a voltage probe to control the working electrode potential. These electrodes are connected to a high input impedance voltmeter and a probe current has to pass through the reference electrode but its absolute value does not exceed 0.01 nA (voltage lower than 1 V and input impedance greater than 10^{11} Ω). As the exchange current of a reference electrode is several orders of magnitude greater than 0.01 nA, the probe current passing through the reference electrode does not change its potential significantly, i.e. this

Table 9.2   Chemical composition of synthetic granitic and clayey groundwater

Element	Groundwater (concentration in mmol/l)	
	Granitic	Clayey
Na^+	117.5	23.57
K^+	0.768	0.11
Ca^{2+}	7.487	8.094
Mg^{2+}	4.115	5.848
Si	0.333	0.071
Al		0.011
Cl^-	112.8	38.46
HCO_3^-	1.536	1.309
SO_4^{2-}	13.53	5.859
F^-	0.157	
pH	8.0 (22°C)	8.5 (23°C)
Resistivity (Ω.cm)	81.6 (22°C)	222 (23°C)

potential can be considered as constant. This set-up can also work in replacing the reference electrode by a corroding electrode as long as the corrosion current is several orders of magnitude greater than the probe current (0.01 nA). Consequently for practical reasons, a corroding electrode can replace the reference electrode. This gives us a three-electrode cell consisting of a counter electrode and two identical corroding electrodes: one is used as a voltage probe, whereas the other is used as a working electrode. This arrangement has two drawbacks: (i) the evolution of the actual corrosion potential is unknown and (ii) stationary voltage does not imply stationary corrosion potential. Nevertheless, this arrangement can be used for corrosion monitoring over long-term experiments (several hundred hours) in pasty mixtures.

9.2.3   Electrochemical autoclave

All of the experiments were carried out in a 316L autoclave under a pressure of 5 MPa. Three temperatures were tested (150, 90 and 50°C). The experiments presented here were performed without oxygen. The air contained in the autoclave gas phase was removed by rough pumping ($\approx$ 1300 Pa). Dissolved oxygen disappeared by natural oxidation of the 316L autoclave, so that the mixture remained oxygen-free for the whole duration of the experiment, which was carried out under an inert atmosphere. Two initial redox potentials of the pasty mixture were tested. One was obtained by introducing a mixture of argon and 0.3% CO_2 in the gas phase. The other was obtained by introducing a gas mixture of argon containing 4% H_2 and 0.3% CO_2. The CO_2 was introduced to buffer the pH of the mixture by equilibrium.

The autoclave itself ($\phi = 100$ mm, height $= 135$ mm) was used as the counter electrode. It was filled with 880 ml of pasty mixture. The carbon steel electrodes were rods ($\phi = 16$ mm, height $= 16$ mm). The upper face of each rod was screwed to the autoclave cap by means of an insulating waterproof connector. This set-up prevents any electrical short circuit between the electrodes and the autoclave. The active surface of each electrode was 10 cm^2. The two identical carbon steel electrodes (working electrode and voltage probe) were immersed in the upper part of the pasty mixture. Rotating propellers in the lower part of the autoclave provided continuous stirring (60 rpm) of the mixture.

9.2.4   Corrosion product analysis

After several hundred hours of immersion in the autoclave, the carbon steel specimens were embedded in pasty clay. This pasty clay was removed by several ultrasonic cleanings in deionised water. Then the plane section of the rods was observed by scanning electron microscopy (SEM). When a thick corrosion product layer was observed, the X-ray diffraction technique was used to try to identify the corrosion products and finally SEM observations of the cross sections were performed.

9.3   Experimental results

9.3.1   Corrosion monitoring

At the beginning of the experiment, during the temperature increase, the absolute voltage between the two identical carbon steel electrodes showed a rise of a few tens

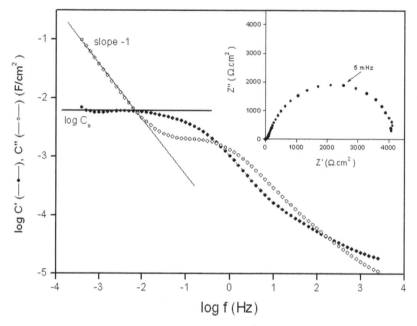

9.1 Capacitance and impedance (insert) spectra obtained on 1050 carbon steel corroded in a de-aerated granitic mixture at 50°C after 330 h

of mV. When the operating temperature was reached, the absolute voltage decreased to zero within the next few hours and remained at this value except in experiments performed at 90°C. For these latter experiments, a long transitory period was observed, during which the absolute voltage first increased and then decreased to zero. An explanation for this transitory period is given in section 9.3.3.

The corrosion monitoring was performed by electrochemical impedance spectroscopy (EIS). In most cases, the impedance spectrum was collected at zero voltage in the 50–10 kHz to 10–0.2 mHz frequency range. Every spectrum exhibited a capacitive loop. For experiments carried out at 50 and 90°C before the transitory period, the spectra obtained were very similar. The capacitive loop was distorted in the high frequency range but exhibited no distortion in the low frequency range (see Fig. 9.1). For the experiments carried out at 150 and 90°C after the transitory period, a similarity between the graphs is also clear, but this time the capacitive loop corresponded to a depressed semicircle (see Fig. 9.2).

An estimation of the transfer resistance (R_t values) was achieved in using the complex capacitance spectrum $C(\omega)$. This spectrum is calculated from the impedance spectrum $Z(\omega)$ after correction for the solution resistance R_e

$$C(\omega) = \frac{1}{i\omega[Z(\omega) - R_e]} = C'(\omega) + iC''(\omega) \qquad [9.1]$$

where $\omega = 2\pi f$ is the pulsation (f the perturbation frequency) and i is the pure imaginary. For a R_t parallel C_o equivalent circuit, the complex capacitance $C(\omega)$ is given by

$$C(\omega) = C_o + \frac{1}{i\omega R_t} \qquad [9.2]$$

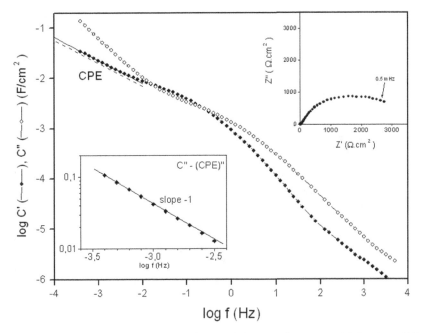

9.2 Capacitance and impedance (upper right insert) spectra obtained on 1050 carbon steel in a de-aerated clayey mixture at 90°C after the transition. Lower left insert shows the variations of $C''(\omega) - \Delta C \sin \dfrac{n\pi}{2} \omega^{-n}$ in a log–log plot

This relationship shows that the imaginary part of the complex capacitance $C''(\omega)$ exhibits a linear relationship in log C''–log f representation. The slope of the straight line is –1. The use of the complex capacitance spectrum allows the determination of the frequency range where the R_t parallel C_o equivalent circuit is valid. Then in using a linear regression, the estimation of the R_t value can be obtained. As can be seen in Fig. 9.1, this procedure has worked very well for experiments carried out at 50 and 90°C before the transitory period.

The impedance spectra obtained at 150 and 90°C after the transitory period exhibited a depressed semicircle as shown in Fig. 9.2. In these cases, the R_t estimation procedure presented above may still stand. Generally, a depressed semicircle is linked to a Constant Phase Element (CPE). Introducing a CPE in relation 9.2 leads to

$$C(\omega) = \Delta C(i\omega)^{-n} + \frac{1}{i\omega R_t}$$ [9.3]

ΔC and exponent n can be obtained from linear regression of log C' vs. log f. In Fig. 9.2, the real part of the CPE is represented by the continuous line, whereas the imaginary part is represented by the dashed line.

The difference between the imaginary part $C''(\omega)$ and $\Delta C \sin \dfrac{n\pi}{2} \omega^{-n}$ gives a remainder, which can be plotted in log–log scale. For all depressed semicircles obtained in this study, a slope of –1 was obtained for the logarithm of the remainder versus log f. The estimation of R_t was then performed as explained above. This procedure has confirmed that the depressed semicircle was linked to a CPE in the low frequency range.

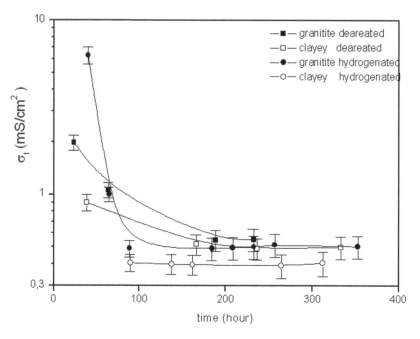

9.3 Time evolution of the transfer conductance σ_t for 1050 carbon steel immersed at 150°C in a mixture of clay and solution (Table 9.2)

In summary, for all impedance spectra obtained in this study, an evaluation for R_t was possible in the low frequency range, even if the imaginary part of the impedance did not tend towards zero (see insert in Fig. 9.2). Furthermore, corrosion monitoring was performed as time evolution of transfer conductance σ_t, which is defined as the inverse of R_t, because σ_t can be regarded as proportional to the corrosion current density.

9.3.2 Time evolution of the transfer conductance at 150°C

The results obtained for time evolution of transfer conductance σ_t at 150°C are presented in Fig. 9.3. For all of the experiments carried out at this temperature, the transfer conductance has decreased over the first hundred hours, indicating some kind of passivation process. All of the spectra exhibited a depressed semicircle (see Fig. 9.2). For longer exposure times, the transfer conductance was stationary and roughly independent of the redox conditions and solution compositions. As will be detailed explained in section 9.3.6, the 'passivation' was due to the formation of a thick corrosion layer.

9.3.3 Time evolution of the transfer conductance at 90°C

The results obtained for time evolution of transfer conductance σ_t at 90°C are presented in Fig. 9.4. For short times, transfer conductance values were high and the impedance spectra exhibited shapes similar to those shown in Fig. 9.1. For longer times, transfer conductance values have decreased by about one decade and depressed semicircles were observed (see Fig. 9.2). The transition between these two situations occurred after 100 to 200 h of exposure, depending on the experiments. Two similar

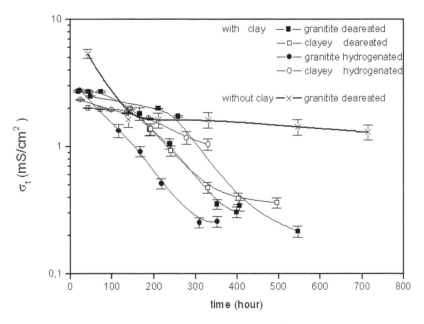

9.4 Time evolution of the transfer conductance σ_t for 1050 carbon steel immersed at 90°C in a mixture of clay and solution (Table 9.2) and in granitic solution without clay

experiments have been carried out in a granitic de-aerated mixture. The time evolutions for the transfer conductance were similar but shifted in time (about 80 h).

As will be shown in section 9.3.6, the transition corresponds to the formation of a stacked bi-layered structure: a calcium siderite $Fe_xCa_{1-x}CO_3$ outer layer on an iron oxide inner layer. A specific experiment has been carried out in granitic solution without 'FoCa' clay. In this case, no transition has occurred even after 700 h. Only iron calcium carbonate dispersed on the steel surface was observed (see Fig. 9.9). The transfer conduction has remained at the value obtained for short time experiments in clay. The shape of the impedance spectra corresponded to those presented in Fig. 9.1. This suggests that for the experimental conditions used in this work, the depressed semicircle was related to a thick corrosion layer.

The two results obtained for the granitic de-aerated water–clay mixture have suggested that the formation of two stacked layers did not always occur at the same time. Therefore it is believed that the formation of these stacked layers did not occur at the same time for the two identical carbon steel electrodes immersed in the autoclave. This could explain the observed non-null voltage between the carbon steel electrodes during the transitory period: the formation headings for stacked layers were not identical at the same exposure time on the two carbon steel electrodes. When the stacked layers were completely formed on both carbon steel electrodes, the corresponding voltage became zero.

9.3.4   Time evolution of the transfer conductance at 50°C

Results obtained for time evolution of transfer conductance σ_t at 50°C are presented in Fig. 9.5. The value for the transfer conductance has decreased with time but this

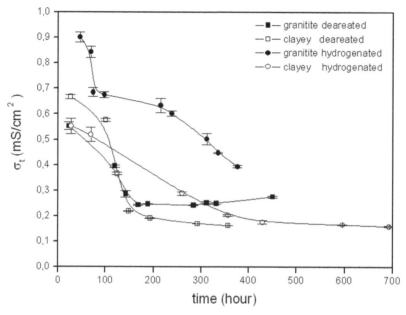

9.5 Time evolution of the transfer conductance σ_t for 1050 carbon steel immersed at 50°C in a mixture of clay and solution (Table 9.2)

decrease was less extensive than for higher temperatures. For longer exposure times, the values obtained for clayey solutions (0.15 mS/cm²) seemed to be lower than those obtained for granitic solutions (0.3–0.4 mS/cm²). It is believed that this trend could be linked to the chloride concentration, which was greater in the granitic solution than in the clayey solution (Table 9.2).

9.3.5 Corrosion products at 150°C

SEM observations and Energy Dispersive X-ray Spectroscopy (EDS) revealed a thick iron oxide layer containing silicon. X-ray diffraction analysis showed relevant peaks at d-spacings around 0.8–0.7, 0.35–0.40 and 0.26–0.30 nm (see Fig. 9.6). Other peaks corresponded to Fe_3C and ferrite α. Assuming that these three peaks corresponded to a single compound, the oxide layer would be Cronstedtite-1M [1] or Greenalite-1M [2]. All are ferro silicate oxyhydroxides. Several formulae have been proposed: $Fe_3^{III}(Fe^{III}Si)O_4(OH)_5$ [1], $(Fe^{II}Fe^{III})(Fe^{III}Si)O_5(OH)_4$ [3], $Fe^{III}Si_2O_5(OH)_4$ [4].

Taken as a whole, they form a 'family of compounds' with similar structure, i.e. stacked large oxygen sub-lattice containing cations. This sheet structure is revealed by a shoulder at $d \approx 0.7$–0.8 nm. Depending on the radii of the cations but also on their localisation in the oxygen sub-lattice, some distortions exist, leading to shifts of some spectral peaks for low values of d-spacing. A full identification could be possible thanks to these extra peaks. Unfortunately, Fe_3C exhibits many major lines in this d-spacing range, preventing any sure identification of this oxide (see Fig. 9.6).

A cross-sectional observation is presented in Fig. 9.7. Steel was covered by a 10 µm thick oxide layer. The outer part of the oxide layer exhibited a dark colour, which is relevant with some doping by silicon.

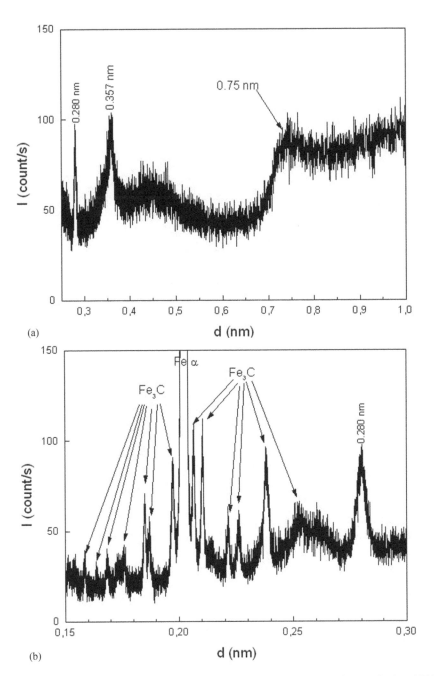

9.6 X-ray diffraction spectrum obtained on 1050 carbon steel corroded at 150°C in a de-aerated clayey mixture after 333 h

9.3.6 Corrosion products at 90°C

SEM observations of plane sections showed that carbon steel specimens were completely covered by a well crystallised layer. EDS analysis indicated that this layer contained Fe, Mn, Ca, C and O. X-ray diffraction indicated that the layers were made

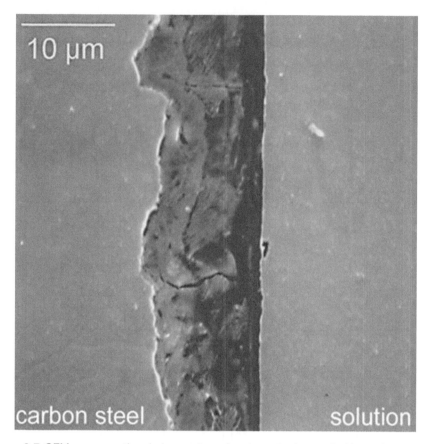

9.7 SEM cross-sectional observation of carbon steel corroded in a mixture of clay and de-aerated granitic solution at 150°C after 232 h

of iron carbonate substituted by calcium $Fe_xCa_{1-x}CO_3$. Substitution of Fe by Ca shifts the peaks toward higher d-spacing values [5]. As substitution of Fe by Mn does not significantly affect the d-spacing values, it seems likely that the layers were rather made of $(Fe, Mn)_x Ca_{1-x}CO_3$ (x ranging from 0.95 to 0.6 and depending on experimental conditions).

SEM cross-sectional observations revealed a stacked bi-layered structure (Fig. 9.8). The outer layer consisted of substituted iron carbonate $(Fe, Mn)_x Ca_{1-x}CO_3$. The inner layer consisted of iron oxide with carbon content (EDS analysis). Detailed analysis of X-ray scattering spectra showed minor peaks due to Fe_3C. It is well known that an iron oxide layer formed on iron in borate buffer has a spinel-like structure [6,7]. Unfortunately, the main peaks for Fe_3O_4, γ-Fe_2O_3 and Fe_3C are located roughly on the same d-spacing values. It was thus impossible to identify the iron oxide from X-ray scattering spectra.

SEM observations of a carbon steel specimen corroded in granitic solution without clay are presented in Fig. 9.9. A dispersed iron calcium carbonate deposit was observed. The shape of the crystallites was similar to those observed in the experiments with clay. This suggested that, in the presence of clay, a continuous layer has been able to form whereas only a scattered deposit has been formed without clay.

inner oxide layer

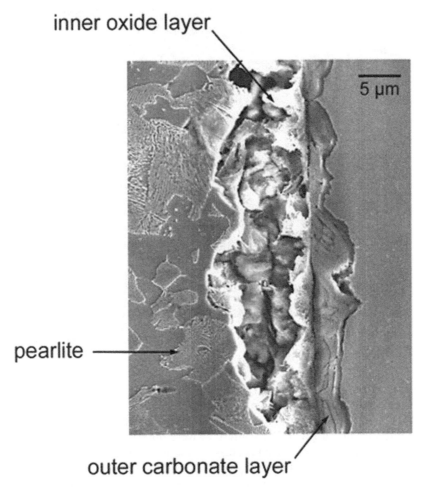

5 µm

pearlite

outer carbonate layer

9.8 SEM cross-sectional observation of carbon steel corroded in a mixture of clay and de-aerated granitic solution at 90°C after 550 h

Spot EDS analysis on areas lying between the deposits exhibited a low intensity X-ray emission for oxygen (0.52 keV), indicating the presence of a thin oxide layer on carbon steel. This suggests that at 90°C, the growth of a thick oxide layer (Fig. 9.8) was linked to the formation of a continuous carbonate outer layer.

9.3.7   Corrosion products at 50°C

SEM observations revealed only the structure of carbon steel, i.e. ferrite and pearlite grains. EDS analysis showed a low intensity X-ray emission for oxygen, indicating that the steel was probably covered by a thin oxide layer.

9.4   Discussion

It is well known that iron corroded in neutral solutions is covered by a thin oxide layer [8–12]. Using ellipsometry, Lukac et al. [13] have measured thicknesses of a few nanometres on iron potentiostatically corroded in borate buffer in the 0–80°C

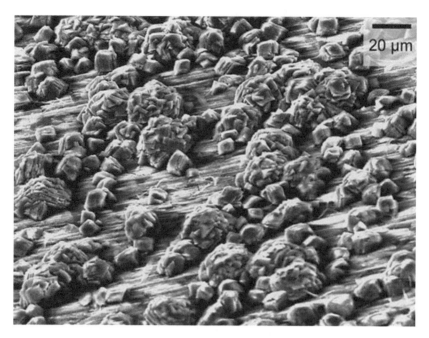

9.9 SEM observation of carbon steel surface corroded in granitic solution without clay at 90°C after 700 h

temperature range. From EDS oxygen analysis, the results obtained at 50°C were in qualitative agreement with these observations.

At 150°C, thick oxide layers were formed. It must be noted in Fig. 9.7 that a straight outer interface and a rough inner interface were observed. This difference in shape has suggested that the interfacial processes were not similar. Using iron reoxidation in $H_2^{18}O$ borate buffer solution and SIMS analysis, Graham et al. [9] have shown that the film oxide growth takes place at the inner interface. So it is believed that the oxide layer growth could take place at the inner interface whereas the oxide layer dissolution could take place at the outer interface. This observation well illustrates the basic assumptions of the *Point Defect Model* [14–16].

At 90°C, a bi-layer stacked structure was observed consisting of an inner oxide layer and an outer carbonate layer. It must be noted in Fig. 9.8 that, once more, the steel/oxide inner layer interface was rough whereas the inner/outer layer interface was straight. Moreover, the outer layer/solution interface was rough. Also, this observation well illustrates the *Point Defect Model* in the case of the formation of a bi-layer passive film [17–19]. Interface shapes suggest that the inner oxide layer has grown at the steel/oxide interface whereas the outer layer has grown at the carbonate/solution interface.

The formation of this bi-layer stacked structure could be explained in the framework of the *Point Defect Model*. This model assumes that the metal is covered by a barrier layer, which grows at the metal/layer interface (inner interface) and dissolves at the layer/solution interface (outer interface). It is also assumed that direct metal oxidation and cationic transport can occur through the barrier layer (vacancy or interstitial hopping transport). On iron, this leads to production of iron cations at the outer interface. In de-aerated or hydrogenated conditions, the predominant oxidation state for iron cations is Fe^{2+}.

Cathodic reduction of water produces hydrogen and hydroxyl anions. The latter can react with bicarbonate to produce CO_3^{2-} ions at the outer interface. So the iron carbonate corrosion product can be formed. As groundwater contains Ca^{2+} and 1050 carbon steel contains Mn, mixed {Fe, Mn, Ca} carbonates have been formed in practice.

A higher corrosion rate leads to higher rates of CO_3^{2-} and Fe^{2+} production at the outer interface. In granitic solution without clay, only dispersed deposits have been observed because diffusion convection in solution was sufficient. In the pasty mixture, diffusion convection was less efficient. The outer interface was more confined than in solutions without clay. Thus a continuous and probably porous carbonate layer has been formed. This outer layer could act as a natural paint in decreasing the dissolution rate of the barrier layer (inner layer). In the *Point Defect Model*, the barrier layer growth is assumed to be independent of the outer interface processes. Therefore the barrier layer (inner layer) could have grown. This could explain why a thick inner layer has been observed. On the other hand, without clay, dispersed deposits did not change the rate of dissolution of the barrier layer significantly. In this case, a thin barrier layer was probably formed.

EIS has clearly given two kinds of spectra. The one depicted in Fig. 9.1 corresponds to a thin oxide layer, whereas the other depicted in Fig. 9.2 corresponds to a thick oxide layer.

It can be noticed on Fig. 9.1 that the real part $C'(\omega)$ seems to tend toward a high frequency limit. This corresponds to the well known universal law for dielectric loss proposed by Jonscher [20,21]

$$C(\omega) = C_\infty + \Delta C(i\omega)^n \qquad [9.4]$$

where C_∞ corresponds to the pure capacitance of the system. The second term corresponds to dielectric loss. In Cole and Cole representation ($C'(\omega)$ vs. $C''(\omega)$) [22], relation 9.4 gives a straight line, which intercepts the real axis at C_∞. For metal covered by an oxide layer, this pure capacitance is given by

$$\frac{1}{C_\infty} = \frac{1}{C_{layer}} + \frac{1}{C_{double\ layer}} = \frac{L}{\chi\chi_o} + \frac{1}{C_{dl}} \qquad [9.5]$$

where L is the thickness of the oxide layer, χ the relative dielectric constant and χ_o the dielectric constant of a vacuum. Considering that χ is 10–30 [23] and that L is a few nanometres, the capacitance layer C_{layer} is of the order of 1–10 $\mu F/cm^2$, that is the same order of magnitude as the double layer capacitance C_{dl}.

In using the same route for thicker layers ($L \approx 10$ μm), a layer capacitance $C_{layer} \approx 1$ nF/cm^2 is obtained. This is much smaller than the double layer capacitance C_{dl}. So C_∞ is practically equal to C_{layer}. This result was observed on the capacitance spectrum depicted in Fig. 9.2. In practice, the intercept with the real axis on a Cole & Cole representation gave zero. This showed that the order of magnitude for C_∞ can be considered as a qualitative measure of the oxide layer thickness, i.e. nanometric layers for $C_\infty \approx 10$ $\mu F/cm^2$, thicker layers for lower values.

9.5   Conclusions

Electrochemical corrosion experiments have been performed on 1050 carbon steel in mixtures of groundwater and 'FoCa' clay, in the temperature range of 50–150°C.

All experiments presented in this chapter concerned mixtures free of oxygen. These experiments can be regarded as long time experiments in comparison with other previously published results because corrosion monitoring was performed over several hundreds of hours. This set of experiments has given an overview of corrosion processes of 1050 carbon steel in the presence of 'FoCa' clay. Other experiments of the same kind are in progress with bentonite MX80 to evaluate the role of clay in corrosion processes of carbon steel.

In the case of 'FoCa' clay, two kinds of results have been observed:

At the lowest temperature (50°C) and at 90°C for short exposure time, 1050 carbon steel was probably covered by a nanometric oxide layer. This case corresponded to a noteworthy complex capacitance spectrum and to a high C_∞ value. The corresponding complex capacitance spectrum consisted of a high frequency range CPE with an exponent of about ½, and an R parallel C equivalent circuit in the low frequency range.

At the highest temperature (150°C) and at 90°C for long exposure times, 1050 carbon steel was covered by a thick micrometric oxide layer directly in contact with steel. This case corresponded to another noteworthy capacitance spectrum and to low C_∞ value. The corresponding complex capacitance spectrum consisted of a high frequency range CPE with an exponent higher than ½ (typically 0.6–0.8) and in a second CPE in parallel with a resistance in the low frequency range.

In both cases, complex capacitance spectra and surface analysis of thick corrosion products have suggested that the corrosion process of this carbon steel could be described at least qualitatively in the framework of the *Point Defect Model*.

Probably the most significant result concerned the long-term behaviour of 1050 carbon steel in the presence of 'FoCa' clay at 90°C. The confinement of the outer interface (layer/pasty mixture) has induced the growth of an outer corrosion product layer, which inhibited the dissolution rate of the oxide layer directly in contact with steel (inner layer). This inhibition has led to the growth of this oxide inner layer.

It would be interesting to move forward in analysing quantitatively the observed complex capacitance spectra in the framework of the *Point Defect Model*.

Acknowledgement

This work has been supported by ANDRA (French National Agency Responsible for the Management of Radioactive Nuclear Waste).

References

1. PCPDFWIN Software version 1.30: JCPDS-International Centre for Diffraction Data Base n°17-0470. Reference to R. Steadman and P. M. Nuttall, *Acta Crystallogr.*, 17 (1964), 404–406.
2. J. J. Papike, *Rev. Geophys.*, 26 (1988), 407–444.
3. R. Steadman and P. M. Nuttall, *Acta Crystallogr.*, 16 (1963), 1–8.
4. PCPDFWIN Software version 1.30: JCPDS-International Centre for Diffraction Data Base n°45-1353.
5. *Landolt-Börnstein, Zahlenwerte und Funktionen aus Naturwissenschaften und Technik, Neue Serie, Gruppe III: Kristall und Festkörperphysik Band 7*, 123–125, Springer-Verlag, Berlin, Heidelberg, New York, 1979.

6. P. Schmuki, S. Virtanen, A. J. Davenport and C. M. Vitus, *J. Electrochem. Soc.*, 143 (1996), 574–582.
7. M. P. Ryan, R. C. Newman and G. E. Thompson, *J. Electrochem. Soc.*, 142 (1995), L177–L179.
8. M. Buchler, P. Schmuki and H. Bohni, *J. Electrochem. Soc.*, 144 (1997), 2307.
9. M. J. Graham, J. A. Bardwell, R. Goetz, D. F. Mitchell and B. MacDougall, *Corros. Sci.*, 31 (1990), 139.
10. N. Sato, T. Noda and K. Kudo, *Electrochim. Acta*, 19 (1974), 471.
11. N. Sato, T. Noda and K. Kudo, *Corros. Sci.*, 10 (1970), 785.
12. M. I. Nagayama and M. Cohen, *J. Electrochem. Soc.*, 109 (1962), 781.
13. C. Lukac, J. B. Lumsden, S. Smialowska and R. W. Staehle, *J. Electrochem. Soc.*, 122 (1975), 1571.
14. C. Y. Chao, L. F. Lin and D. D. MacDonald, *J. Electrochem. Soc.*, 128 (1981), 1187.
15. L. Zhang and D. D. MacDonald, *Electrochim. Acta*, 43 (1998), 679.
16. D. D. MacDonald, *Pure Appl. Chem.*, 71 (1999), 951.
17. D. D. MacDonald and M. Urquidi-MacDonald, *J. Electrochem. Soc.*, 137 (1990), 2395.
18. D. D. MacDonald, S. R. Biaggio and H. Song, *J. Electrochem. Soc.*, 139 (1992), 170.
19. D. D. MacDonald, *J. Electrochem. Soc.*, 139 (1992), 3434.
20. A. K. Jonscher, in *Advances in Research and Development*, 217, ed. G. Hass and M. H. Francombe, Vol. 11, Academic Press, New York, 1980.
21. A. K. Jonscher, *Physics of Dielectric*, 22, Inst. Phys. Conf. Ser. No. 58, 1980.
22. D. M. Smyth, in *Oxides and Oxide Films*, 110, ed. J. W. Diggle, Vol. 2, Marcel Dekker, Inc, New York, 1973.
23. G. V. Samsonov (ed.), *The Oxide Handbook*, 211, 2nd edn. IFI, New York, Washington, London, 1982.

10

Corrosion monitoring applications in nuclear power plants – a review

Shunsuke Uchida

Nuclear Science and Engineering Directorate, Japan Atomic Energy Agency, 2-4, Shirane, Shirakata, Tokai, Ibaraki, Japan

uchida.shunsuke@jaea.go.jp

10.1   Introduction

Water chemistry is not only one of the most important parameters to maintain plant reliability, but it is also one of the most important indices to measure plant operational conditions. In order to understand plant conditions, *in-situ* measurements of local conditions, e.g. in-core corrosive conditions, are desired but most water chemistry data have been obtained with probe type sensors applied in room temperature water. Some mismatching has been reported between the information required to understand plant conditions at elevated temperatures and the water chemistry data measured at room temperature [1]. From the viewpoint of water chemists, procedures for bridging the gaps as well as providing quality assurance of water chemistry data acquisition are essential subjects. The major measures to bridge the gaps are of two types: theoretical approaches and direct determinations by applying high-temperature water chemistry sensors.

In this chapter, data acquisition and processing systems in boiling water reactor (BWR) and pressurised water reactor (PWR) plants are briefly introduced, the necessity of adequate water chemistry control to obtain plant reliability and safety is presented, and procedures to bridge the gaps between the measured water chemistry data and the information needed to understand real interactions between materials and water in the primary cooling systems are described. The latest developments and applications of high-temperature water chemistry sensors are also briefly reviewed and major proposals for their application in operating plants and their future possibilities are discussed.

10.2   Requirements for in-line water chemistry sensors

The most important roles of cooling water in light water reactors (LWRs) are energy transport and neutron moderation. At the same time, the resulting high-temperature water causes corrosion of structural materials, which leads to adverse effects in the plants, e.g. increasing shutdown radiation, generating defects in materials of major components and fuel claddings, and increasing the volume of radwaste sources. In Fig. 10.1, the major roles of cooling water in BWRs and PWRs are summarised [2–4].

In order to control the adverse effects shown in Fig. 10.1, it is essential to understand corrosion behaviours of structural materials and then to control them.

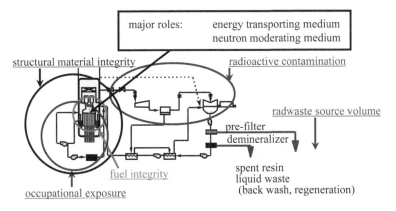

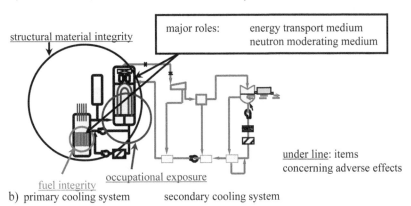

10.1 Major roles and adverse effects of cooling water of nuclear power plants; a) BWR primary cooling water; b) PWR primary and secondary cooling water

Corrosion behaviours are greatly affected by the combinations of water quality and materials. In order to minimise the adverse effects, optimal water chemistry control has been proposed as shown in Fig. 10.2 [4].

It is important to improve water chemistry by evaluating the possible effects of small changes in water chemistry on fuel cladding and structural materials after a long incubation time, as well as to consider the possibility for changes in materials which cause long-term degradation of water chemistry [5].

In order to evaluate changes in water chemistry, *in-situ* measurement of water quality is required. Water chemistry conditions should be monitored without any change of the chemical conditions. They should be monitored continuously with high accuracy and high reliability to be prepared for a quick response if there are any anomalous conditions, and without any adverse effect on cooling systems, even if the sensors are damaged.

Recent computer aided medical inspection systems are supported by automatic analysis systems for blood tests, urine analyses and electrocardiographs. In particular, the blood furnishes information on major organs as it circulates through them. Anomalous conditions in the organs affect the blood properties and *vice versa*. By analogy, in a BWR, the primary coolant has almost the same role as blood in the

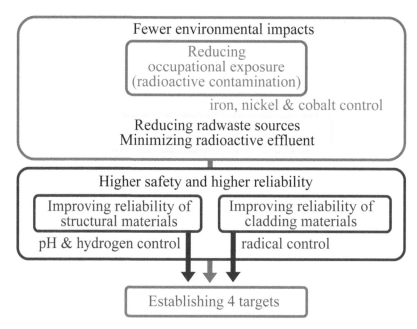

10.2 Optimal water chemistry control (BWR and PWR)

human body [5]. The coolant circulates through the major components in the primary system, such as the core, recirculation pumps, turbines and water polishing systems, each of which can correspond to a major organ. Suitable and timely plant renewal is required to maintain sufficient reliability for ageing BWRs, just as medical treatments and preventive measures are needed for age-related adult diseases. In addition to annual inspection, power anomaly diagnosis can prepare accurate information on structural materials as well as plant functions, which allow suitable and timely preventive maintenance for ageing plants.

Cooling systems, major components and water chemistry differ in BWR and PWR plants. Procedures to measure water chemistry are also different in both plants. Locations for in-line monitoring and sampling in BWRs and PWRs are shown in Fig. 10.3 [1]. For more detailed measurements, cooled and depressurised water is sampled and brought to a chemical laboratory in the plants for chemical and radioactive nuclide analysis.

Normal in-line water chemistry sensors, i.e. pH, conductivity, oxygen concentration [O_2], and hydrogen concentration [H_2], give important information to understand the corrosive conditions of the primary coolant. However, the features of water chemistry measured with in-line sensors are those after completion of a cooling procedure and some changes in water chemistry data are considered. Conductivities, pH and [O_2] are measured at room temperature and their effects on the materials should be considered by extrapolating them to values at elevated temperatures. Major methods to obtain water chemistry and data of concern are summarised in Table 10.1.

One of the most important radiolytic species in BWR cooling water, H_2O_2, which cannot be measured in the sampled water, should be determined by the theoretical water radiolysis model along with O_2 and H_2 at the location of interest.

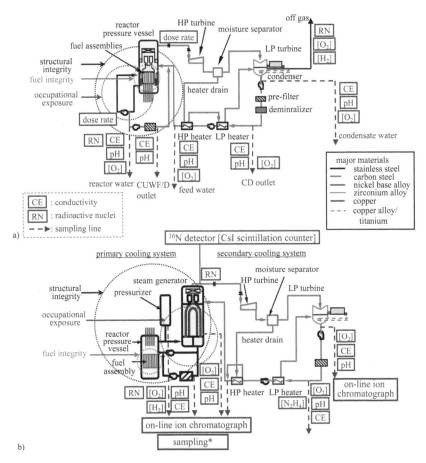

10.3 Major subjects related to materials in cooling systems of nuclear power plants; a) BWR; b) PWR

The oxygen concentration distribution along the PWR feed water flow path is calculated based on the oxygen hydrazine reaction. $[O_2]$ in the cooling water can be confirmed by measuring it in the sampled water taken from the complex sampling line (Fig. 10.3b) but the N_2H_4–O_2 reaction also occurs in the sampling line. The gap between the sampled $[O_2]$ and the measured value should be bridged to evaluate the actual concentration. *In-situ* measurement of $[O_2]$ in high-temperature water is desired. Electrochemical corrosion potential (ECP) is one of the candidates for high-temperature water chemistry sensors for corrosive conditions determined by O_2 and N_2H_4, where ECP is expressed as a function of $[O_2]$ and $[N_2H_4]$ [6,7].

Major gaps between the desired information to understand the phenomena and measured data are shown in Table 10.2 with measures to bridge these gaps. Some mismatching between the information desired to understand plant conditions at elevated temperature and the water chemistry data measured at room temperature have been reported [1]. From the viewpoint of water chemists, procedures for bridging the gaps as well as providing quality assurance of water chemistry data acquisition are essential subjects. The major measures to bridge the gaps are of two types: theoretical approaches and direct determinations by applying high-temperature water chemistry sensors.

Table 10.1   Major methods to obtain water chemistry data (BWR and PWR)

Item	Method	Sampled volume	Data transfer	
			BWR	PWR
Temperature	Thermocouple		on (HT)	on (HT)
Pressure	Diaphragm type pressure gauge		on (HT)	on (HT)
Flow rate	Orifice type pressure gauge		on (HT)	on (HT)
γ-ray spectroscopy	Ge(Li)		on (HT*)	on (HT*)
Detection of β emitters	LSC:		off	on (HT*)
Metallic impurities	XFS	10–100 l	off	off
Others	TOC		off	off
Backup	AAS, SP, ICP, ICP-MS, IC, NaI(Tl)		off	off
Conductivity			on (RT)	on (RT)
pH			on (RT)	on (RT)
[O2]			on (RT)	on (RT)
[H2]			on (RT)	on (RT)
<ain steam radiation			on (RT*)	on (RT*)
Off gas nuclide			on (RT*)	–

on: on line sensors. off: off line sensors, (HT): high temperature sensors, (HT*): room temperature sensors for high temperature samples, (RT): room temperature sensors

Ge(Li) : Ge(Li) semiconductor detector
TOC: total organic carbon detector
ICP: induction coupled plasma spectroscopy
NaI(Tl) : NaI(Tl) scintillation counter

LSC: liquid scintillation counter
AAS: atom absorption spectrometry
IC: ion chromatograph

XFS: X-ray fluorescence spectrometry
SP: spectrophotometer

ICP-MS: ICP mass spectroscopy

10.3   High-temperature water chemistry sensors

High-temperature water chemistry sensors are divided into those determining water chemistry directly and those determining corrosive conditions indirectly by analysing the interaction between water and materials. Major high-temperature water chemistry sensors are listed in Table 10.3 [8–10].

The purposes of high-temperature water chemistry sensors are also divided into three categories:

Category 1:   Direct determination of water chemistry index at elevated temperature instead of theoretical extrapolation from the data obtained at room temperature. Changes in data at room temperature and elevated temperature are estimated but the extrapolation procedures might contain certain uncertainties (pH, conductivity).

Category 2:   Impossible to determine water chemistry index at room temperature. Some chemical species disappear due to thermal decomposition during the cooling process ([H_2O_2] in the BWR water) and chemical reactions in the sampling line ([O_2] in the PWR feed water).

Table 10.2 Gaps between desired information and measured data

Desired information to understand phenomena	Measured WC data in plants	Major measures to bridge the gaps
Corrosive conditions $[H_2O_2, O_2, H_2]$	• Measured $[O_2, H_2]$	• Theoretical models for water radiolysis • HT O_2 sensors • ECP sensors
Crack propagation rate	• Crack growth rate under simulated conditions	• HT crack growth rate sensors • Theoretical & empirical models for crack propagation rate
High temperature pH	• pH of cooled water	• Theoretical evaluation • HT pH sensors
Properties of oxide film on sampled specimens	• Characterisation of oxide film	• Theoretical oxidation models • HT impedance sensors

Category 3:   Direct interaction of water and materials at elevated temperature (ECP, corrosion rate, crack growth).

International collaborative programmes on high-temperature water chemistry sensors have been carried out by the International Atomic Energy Agency (IAEA) (WACOLOIN [8]: 1986–1991; WACOL [9]: 1992–2000; and DAWAC [10]: 2001–2005). From literature surveys and international exchange of information and experience with high-temperature sensors, it was found that many kinds of high-temperature sensors had been developed for direct measurement of water quality, and some of them had been successfully applied in laboratory tests, but only a few had been applied to operating power plants [1].

There are three requirements for application of high-temperature sensors in operating plants:

1) High accuracy: direct application without a cooling procedure to determine the water chemistry index directly.
2) Quick response: direct connection between the sensor and data acquisition system.
3) High reliability and safety: easy calibration, maintenance-free use for a long plant operation period (1 to 1.5 years) and installation downstream from cooling systems.

Figure 10.4 shows the temperature-dependent conductivity and pH for pure water. For pure water, it is easy to extrapolate to values for elevated temperature from measured values at room temperature by using the temperature-dependent dissociation constant of water [11]. However, there are some amounts of coexisting chemical species in the cooling water, which might be added for chemical control of pH and oxidant reduction in PWRs and mixed as chemical impurities in BWRs.

Evaluation of pH and conductivity at elevated temperature from those values measured at room temperature has required some time delay due to the whole

Table 10.3 High-temperature water chemistry sensors

Measured items	Detector	Principle	Applied	Ref.
1) Water chemistry				
a) $[O_2]$	pressure balance type	membrane + polarography	LAB	[16]
	ECP detector	ECP-$[O_2]$ relationship	LAB	[1, 9]
b) $[H_2]$	pressure balance type	membrane + polarography	LAB	[16]
	Pd wire type H_2 detector	H_2 absorption + ER	LAB	[17]
c) $[H_2O_2]$	Sensor array	a couple of ECP and FDCI	LAB	[18]
d) Conductivity	coupled/triple electrodes	complex impedance	LAB	[19]
e) pH	ZrO_2 type pH sensor	ion electrode	LAB	[20]
	TiO_2 type pH sensor	flat band potential	LAB	[21]
2) interaction between water and materials				
a) ECP	external (Ag/AgCl)	ion electrode	NPP	[1]
	internal (Pt)	ion electrode	NPP	[22]
	internal (Ag/AgCl)	ion electrode	NPP	[23]
	internal (Fe/Fe_3O_4)	ion electrode	NPP	[24]
	internal (Ni/NiO)	ion electrode	LAB	[25]
b) Corrosion rate	DC current measurement	ER change due to thinning	LAB	[9]
	DC corrosion detector	Tafel plot	LAB	[9]
	AC corrosion detector	Board diagram	LAB	[9]
c) Oxide film properties	CDE-arrangement detector	contact electric resistance measurement	LAB	[26]
				[27]
	FDCI sensor	complex impedance	LAB	[18]
d) Crack initiation	electrochemical potential detector	potential noise analysis	LAB	[28]
				[29]
e) Crack growth	CT specimen	potential drop measurement	NPP	[30]
	DCB type	potential drop measurement	NPP	[31]

Lab: used only in laboratory experiments
NPP: applied in operating nuclear power plants
ER: electric resistance   FDCI: frequency dependent complex impedance
CT: compact tension   DCB: double cantilever beam
CDE: controlled distance electrochemistry

data arrangement for additional chemical species and the extrapolation calculation and this has often caused undesirable uncertainty mainly due to the influence of coexisting species. Applications of high-temperature pH and conductivity sensors are expected to overcome these difficulties. In order to evaluate corrosive conditions in BWR primary cooling systems, non-linear rate equations are calculated to obtain the concentration distributions of radiolytic species throughout the primary coolant

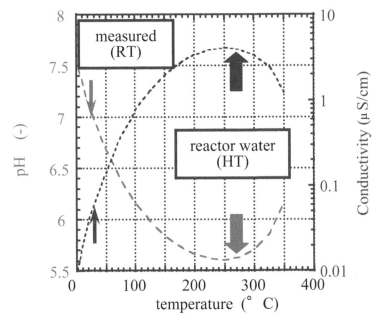

10.4 Conductivity and pH at room temperature and elevated temperature for pure water

[12]. The gap between the measured ECP and effective oxygen concentration ($[O_2]_{eff}=[O_2]+0.5[H_2O_2]$) (Fig. 10.5a) [13] is explained by the contribution of H_2O_2 to the ECP especially at high H_2 injection rate (Fig. 10.5b) [14].

Unfortunately, $[H_2O_2]$ is unstable in high-temperature water and most of it disappears during the sampling process, so it is not possible to determine its concentration in the sampled water. The procedures to determine $[H_2O_2]$ are either by application of the theoretical radiolysis model or direct or indirect measurement at elevated temperature.

As $[H_2O_2]$ decreased from 100 to 5 ppb, the radii of the low frequency semicircles on the Cole–Cole plots of frequency-dependent complex impedance (FDCI) increased

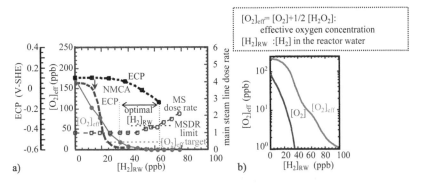

10.5 Effects of H_2 injection in BWR plants; a) measured in plants; b) calculated for plants

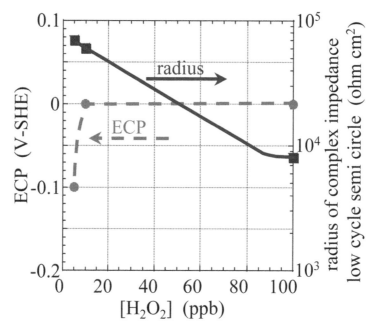

10.6 Relationships between [H₂O₂], ECP and radius of low frequency semicircle on Cole–Cole plots of frequency-dependent complex impedance

continuously, while ECP values remained at the same level from 100 to 10 ppb and then decreased a little to 5 ppb (Fig. 10.6) [15]. The increasing radii meant increasing resistance of oxide dissolution. Both dependences of ECP and FDCI semi-circle radii on [H₂O₂] were not affected by co-existent [O₂] for the same level of oxidant concentration.

As a result of water radiolysis evaluation for BWR water, it was assumed that 100 ppb H₂O₂ co-existed with 200 ppb O₂ under normal water chemistry (NWC), while 10 ppb H₂O₂ existed without a measurable concentration of O₂ under hydrogen water chemistry (HWC). This meant that the corrosive conditions of HWC were the same as those of NWC from the viewpoint of ECP, while [H₂O₂] varied over a wide range, which might affect the corrosion behaviour of structural materials.

Extrapolation of ECP, corrosion rate and crack growth rate from the data obtained at room temperature causes a very large uncertainty due to insufficient databases which could support the extrapolation.

10.4   Experience with high-temperature water chemistry sensors in operating plants

High-temperature sensors used in operating plants and for the in-pile loop of experimental reactors are listed in Table 10.4 [1]. Most of them were sensors for structural material integrity tests. High-temperature reference electrodes for ECP measurements and *in-situ* potential drop detectors for crack propagation measurements, attached to compact tension test specimens, were applied for benchmark tests of HWC in operating BWR plants.

Table 10.4 High-temperature sensors for water chemistry monitoring in plants and in-pile loops

Sensors	Plants	Sensor location	Ref.
ECP(Ag/AgCl)	BWR	AC, LPRMH, BD	[23]
ECP(Cu/CuO$_2$)	BWR	AC	[23]
ECP(Fe/Fe$_3$O$_4$)	BWR	BD	[24]
ECP(Pt)	BWR	LPRMH, BD	[24]
	PWR	core top	[22]
	PWR	FW	[32]
ECP(Pd-hydride)	Halden reactor	in-pile loop	[33]
SSRT	BWR	AC	[1]
CT	BWR	AC	[1]
	Halden reactor	in-pile loop	[34]

ECP: electrochemical corrosion potential
SSRT: slow strain rate test   CT: compact tension
AC: autoclave in reactor water cleanup line   BD: bottom drain
LPRM: local power range monitor housing
FW: feed water

In order to obtain a reductive environment and to mitigate secondary side corrosion of steam generator (SG) tubing, the optimum hydrazine content in the secondary system of the PWR should be decided based on ECP measurements. However, ECP measurements have been made in only a very few units. Once the optimum hydrazine condition is defined, the plant staff only routinely monitor hydrazine and ECP measurements are usually terminated. Instead of direct ECP measurements, a combined approach of concentration measurements of anions and cations by ion chromatography and empirical calculations based on crevice concentration factors and pH evaluation has been successfully applied to determine the corrosive conditions at the tubing and the crevice between the tubing and the supporting plate.

It has been reported that high-temperature water chemistry sensors, i.e. pH, conductivity and [O$_2$] sensors, provided valuable data in laboratory tests. However, a decision on their application in operating plants has not yet been made. In order to enhance the application of high-temperature water chemistry sensors in operating power plants, the following barriers to their applications should be overcome:

1) Clear presentation of their necessity and benefits.
2) Improvement of their reliability during operation, focusing on maintaining sufficient integrity of the sensors themselves and establishing periodical calibration procedures.
3) Mitigation of safety issues regarding defective sensors, focusing on suitable procedures to avoid loose parts and chemical contamination.
4) Completion of suitable case studies for application of plant diagnosis systems based on high-temperature sensors.

The only purpose of operating nuclear power plants is the safe, stable and economical supply of electricity. Unfortunately, the reliability of high-temperature water chemistry sensors is not sufficiently high. Problems with high-temperature water chemistry sensors are divided into two: one is out of signal due to functional defects and the other is the result of loose parts due to mechanical damage. The former can be tolerated losing target information without any real damage, while the latter should be avoided absolutely so as to circumvent plant shutdown. It is important to evaluate the balances between the benefits of sensors, e.g. having information on water chemistry and corrosion of materials, and their risks caused by mechanical damage.

It is recognised that ECP sensors are hard to replace for the evaluation of corrosive conditions, which are determined by unstable oxidants present in the BWR primary coolant and PWR secondary coolant. Their benefits exceed their risks. However, pH, conductivity and other values can be extrapolated from the data measured at room temperature even though uncertainties caused by the extrapolation are often unacceptably large. In these cases, their benefits balance with the risks and then their application in operating plants is debatable.

As a result of improving their reliability and enhancing their importance in data applications, many kinds of high-temperature sensor can be applied in operating plants.

Much experience with the application of sensors in laboratory tests and improvements in their reliability and handling hold the promise of opportunities for their application not only in nuclear plants but also in thermal power plants. The existence of unstable radiolytic species under irradiation makes the application of high-temperature sensors necessary, but at the same time, irradiation damages their reliability in nuclear plants. Complicated chemical treatment makes their application necessary in thermal power plants, but for both environments, high-temperature conditions complicate the selection of sensor materials.

10.5   Conclusions

1) High-temperature water chemistry sensors have been developed for plant applications, but, so far, only ECP sensors have been applied in operating power plants.
2) In order to apply high-temperature water chemistry sensors in operating power plants:

   - their reliabilities should be greatly improved;
   - their application needs should be promoted; and
   - case studies should be demonstrated for applications in plant diagnosis systems.

Acknowledgements

The author expresses his thanks to the members of the nuclear committee of the IAPWS for their enthusiastic assistance.

References

1. S. Uchida, K. Ishigure, H. Takamatsu, H. Takiguchi, M. Nakagami and M. Matsui, 'Water chemistry data acquisition, processing, evaluation and diagnosis systems for nuclear power reactors', in *Proc. 14th Int. Conf. on Properties of Water and Steam* (Kyoto, Japan). International Association for Properties of Water and Steam, Maruzen Co. Ltd, 2004, 551–558.

2. H. Takiguchi, H. Takamatsu, S. Uchida, K. Ishigure, M. Nakagami and M. Matsui, *J. Nucl. Sci. Technol.*, 41 (2004), 214–225.

3. S. Uchida, M. Miki, T. Masuda, H. Nagao and K. Otoha, *J. Nucl. Sci. Technol.*, 24 (1987), 593–600.

4. K. Otoha, N. Uetake and S. Uchida, *J. Nucl. Sci. Technol.*, 34 (1997), 948–956.

5. S. Uchida, Y. Asakura, M. Kitamura and K. Ohsumi, *J. Nucl. Sci. Technol.*, 23 (1986), 233–243.

6. M. Ullberg, P. Andersson, K. Fruzzetti and H. Takiguchi, 'Modeling the oxygen-hydrazine reaction using electrochemical kinetics', in *Proc. 13th Int. Conf. on Environmental Degradation of Materials in Nuclear Power Systems – Water Reactors* (Whistler, BC, Canada, 19–23 August 2007). CNS (CD-ROM).

7. H. Takiguchi, E. Kadoi and M. Ullberg, 'Study on application of oxygenated water chemistry for suppression of flow assisted corrosion in secondary system of PWRs'. in *Proc. 14th Int. Conf. on Properties of Water and Steam* (Kyoto, Japan). International Association for Properties of Water and Steam, Maruzen Co. Ltd, 2004, 531–538.

8. International Atomic Energy Agency, *Coolant Technology of Water Cooled Reactors*, IAEA-TECDOC-667. IAEA, 1992.

9. International Atomic Energy Agency, *High Temperature On-Line Monitoring of Water Chemistry and Corrosion Control in Water Cooled Power Reactors*, IAEA-TECDOC-1303. IAEA, 2002.

10. International Atomic Energy Agency, *Data Processing Technologies and Diagnostics for Water Chemistry and Corrosion in Nuclear Power Plants*, IAEA-TECDOC-1505. IAEA, 2006.

11. A. S. Quist and W. L. Marshall, *J. Phys. Chem.*, 69 (1965), 2984–2987.

12. E. Ibe and S. Uchida, *Nucl. Sci. Eng.,* 89 (1985), 330–350.

13. S. Uchida, E. Ibe, K. Nakata, M. Fuse, K. Ohsumi and Y. Takashima, *Nucl. Technol.*, 110 (1995), 250–257.

14. Y. Wada, S. Uchida., M. Nakamura and K. Akamine, *J. Nucl. Sci. Technol.*, 36 (1999), 169–178.

15. S. Uchida, T. Satoh, K. Iinuma and Y. Satoh, 'Effects of hydrogen peroxide on intergranular stress corrosion cracking of stainless steel', in *Proc. Int. Conf. on Water Chemistry of Nuclear Reactor Systems* (San Francisco, CA, USA, 11–14 October 2004). EPRI, 518–527 (CD).

16. N. Nakayama and S. Uchida, *J. Nucl. Sci. Technol.*, 21 (1984), 476–483.

17. C. Liu and D. D. Macdonald, *J. Supercritical Fluid*, 8 (1995), 263–270.

18. S. Uchida, T. Satoh, N. Kakinuma, T. Miyazawa, Y. Satoh and K. Makela, *ECS Trans.,* 2 (25) (2007), 39–50.

19. Y. Asakura, M. Nagase, M. Sakagami and S. Uchida, *J. Electrochem. Soc.*, 136 (1989), 3309–3313.

20. K. Tachibana, *Mater. Jpn.*, 34 (1995), 1227–1232 (in Japanese).

21. N. Hara and K. Sugimoto, *J. Electrochem. Soc.*, 137 (1990), 2517–2523.

22. N. Nagata, 'In core measurement of electrochemical corrosion potential in Tsuruga-3', in Int. Workshop on Optimization of Dissolved Hydrogen Content in PWR Primary Coolant (Sendia, Japan, 18–19 July 2007), paper 16.

23. S. Ashida, J. Takagi and N. Ichikawa, 'First experience of hydrogen water chemistry at Japanese BWRs', in *Proc. Water Chemistry of Nuclear Reactor Systems*, 6 (Bournemouth, UK, 12–15 October 1992). British Nuclear Energy Society, 103–110.

24. H. Takiguchi, S. Sekiguchi, A. Abe, K. Akamine, M. Sakai, Y. Wada and S. Uchida, *J. Nucl. Sci. Technol.*, 36 (1999), 179–188.

25. R.-W. Bosch, Z. Kerner, G. Nagy, D. Feron, M. Navas, W. Bogaerts, D. Karnik, T. Dorsch, R. Kilian, M. Ullberg, A. Molander, K. Makela and T. Saario, 'LIRES: European sponsored research project to develop light water reactor reference electrode', in EUROCORR (Budapest, Hungary 28 September–2 October 2003).

26. M. Bojinov, 'Development of electrochemical techniques to study oxide films on construction materials in high temperature water', in *Proc. JAIF Int. Conf. on Water Chemistry in Nuclear Power Plants* (Kashiwazaki, Japan, October 1998). Japan Atomic Industry Forum, Inc., 111–116.

27. M. Bojinov, T. Laitinen, K. Makela and T. Saario, *J. Electrochem. Soc.*, 148 (2001), B243–B250.

28. J. Hickling, D. F. Taylor and P. L. Andresen, *Mater. Corros.*, 49 (1998), 651–658.

29. C. Liu, D. D. Macdonald, E. Medina, J. J. Villa and J. M. Bueno, *Corrosion*, 50 (1994), 687–694.

30. W. R. Catlin, D. C. Lord, T. A. Prater and L. F. Coffin, 'The reversing D-C electric potential method', in *Automated Test Methods for Fracture and Fatigue Crack Growth*, ed. W. H. Cullen, R. W. Landgraf, L. R. Kaisand and J. H. Underwood. American Society for Testing and Materials, Philadelphia, PA. *ASTM STP*, 887 (1985), 67.

31. D. Weinstein, 'Real time in reactor monitoring of double cantilever beam crack growth sensors', in *Proc. 6th Int. Symp. on Environmental Degradation of Materials in Nuclear Power Systems – Water Reactors*, 1993, 645–650.

32. A. Molander, B. Rosborg, P.-O. Anderson and L. Bjornkvist, 'Electrochemical measurements in secondary system of Ringhals 3 PWR', in *Proc. Water Chemistry of Nuclear Reactor Systems*, 6, Vol. 2 (Bournemouth, UK, 12–15 October 1992). British Nuclear Energy Society, 228–234.

33. K. Makela and J. K. Peterssen, *Current Status Pd Reference Electrode Work*, Halden Work Report, HWR-449, April 1996, Norway.

34. T. M. Karlsen and E. Hauso, 'Water chemistry and stress intensity effects on the cracking behavior of irradiated austenitic stainless steel', in *Proc. 9th Int. Symp on Environmental Degradation of Materials in Nuclear Power Systems – Water reactors* (Newport Beach, CA, 1–5 August 1999). TMS.

11

In-core corrosion monitoring in the Halden test reactor

Peter Bennett and Torill Karlsen

OECD Halden Reactor Project, PO Box 173, 1751 Halden, Norway

peter.bennett@hrp.no

11.1   Introduction

As power plants age, data on the materials properties of in-core components with high irradiation doses are required for assessments of plant lifetimes, which form the basis for licenses for further operation. Furthermore, as reactor operators implement longer fuel cycles and power upratings, the corrosion behaviour of the alloys used for in-core structures and fuel rod cladding must be studied under the new operating conditions. New alloys offering improved performance are under constant development; while in conjunction, changes in water chemistry conditions can be implemented to mitigate specific corrosion phenomena. An understanding of the mechanisms that lead to material degradation can help to improve and optimise materials properties and to define and qualify water chemistry changes. Studies should be performed under representative conditions to ensure the validity of the data.

The OECD Halden Reactor Project is a joint undertaking of national organisations in 18 countries sponsoring a jointly financed research programme under the auspices of the OECD Nuclear Energy Agency. The Halden Boiling Water Reactor (HBWR) has been in operation since 1958. It is a test reactor with a maximum power of 20 MW and is cooled and moderated by boiling heavy water, with a normal operating temperature of 235°C and a pressure of 34 bar.

The reactor has been predominantly used for investigation of important fuel properties (temperature, pressure, fuel pellet dimensional changes, etc.), where the overall aim has been to assess the performance of current and advanced fuel under normal, abnormal and accident conditions. The results are used for fuel behaviour model development and verification, and in safety analyses [1].

In the past 20 years, increasing emphasis has been placed on corrosion testing, both of in-core structural materials and fuel claddings [2,3]. These tests require representative light water reactor (LWR) conditions, which are achieved by housing the test rigs in pressure flasks that are positioned in fuel channels in the reactor and connected to dedicated water loops, in which boiling water reactor (BWR), pressurised water reactor (PWR) or pressurised heavy water reactor (PHWR) conditions are simulated.

Understanding of the in-core corrosion behaviour of fuel or reactor materials can be greatly improved by on-line measurements during power operation. The Halden Project has performed in-pile measurements for a period of over 35 years, beginning with fuel temperature measurements using thermocouples and use of differential transformers for measurement of fuel pellet or cladding dimensional changes and internal rod pressure [1]. Experience gained over this period has been applied to on-line instrumentation for use in materials tests [4]. This paper gives details of the techniques currently used to provide on-line, in-core corrosion monitoring data, and illustrative results are presented.

11.2   Description of test loops

Test rigs for corrosion studies are installed in pressure flasks, which are constructed from stainless steel and/or Zircaloy and connected to a loop system. The flasks are often surrounded by highly enriched (typically 13–15% enrichment) booster fuel rods, for increasing the fast neutron flux to levels typical of those in commercial LWRs.

A simplified view of a loop system is shown in Fig. 11.1. Specific BWR, PWR or PHWR conditions can be simulated by varying the pressure and temperature of the coolant and the concentrations of dissolved additives and gases. Each loop consists of three main sections: the loop itself, the purification system and the sampling system. Other components include the main circulation pump, heaters/coolers and valves. Main operating parameters, such as temperature, pressure and flow rate are automatically controlled within specified limits using Programmable Logic Control (PLC).

The loops are constructed from 316L stainless steel and have a volume (including the purification system) of between 60 and 120 litres. The loop circulation pumps have capacities from 100 litres to 10 tons per hour, and electric heating is used to ensure that the desired temperature can be maintained in the test section.

Chemistry conditions in the loop are controlled by additives and by the purification plant. Impurities are removed by lithium, boron or mixed bed ion exchange units, or by cartridge filters. Under normal operation, the flow through the purification plant is approximately one to two loop volumes per hour.

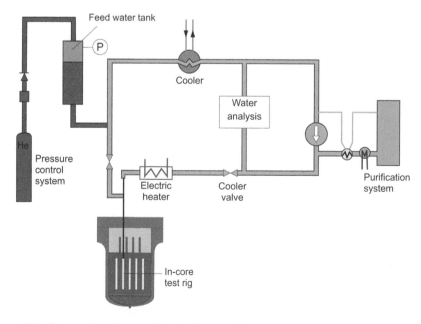

11.1 Schematic diagram of water loop system

The sampling system is used to obtain representative water samples from the loop, to monitor continuously conductivity and dissolved oxygen and hydrogen concentrations in the coolant and to provide a facility for the controlled injection of impurities into the loop. The flow rate through the sample loop is normally in the range from 20 to 50 l h^{-1}. The full loop pressure is maintained in the sampling line; however, the coolant temperature is quickly reduced to room temperature to ensure that the samples are representative.

Grab samples of the coolant are analysed using inductively coupled plasma mass spectrometry (ICPMS) to determine the concentrations of soluble transition metal cations, and by capillary electrophoresis for analyses of dissolved anions. Dissolved oxygen and hydrogen gases are monitored by Orbisphere detectors placed in series in the sampling system. Filter packs, mounted in parallel before the gas analysers, are used to determine integrated concentrations in the coolant of both soluble and insoluble corrosion products and radioactive transition metal species. Approximately 20 litres of coolant are concentrated onto the packs, which contain a Millipore filter (to retain insoluble material) and ion exchange membranes (to retain soluble cationic and anionic species). The filters are subsequently analysed using X-ray fluorescence spectroscopy (XRF) and gamma spectrometry.

11.3   On-line crack growth rate measurements

Irradiation-assisted stress corrosion cracking (IASCC) is used to describe the degradation phenomenon caused by the synergistic effects of neutron radiation, an aggressive aqueous environment and stress/strain on core components, and is widely recognised as one of the most important issues affecting the long-term integrity of LWRs as they age.

In the in-pile crack growth rate investigations being performed in the Halden Reactor, the effects of several of the key factors (stress intensity level, corrosion potential, temperature and yield strength) influencing the stress corrosion cracking (SCC) susceptibility of irradiated core component materials are studied [5,6]. Small compact tension (C(T)) specimens, which are instrumented for crack propagation monitoring with the direct current potential drop (DCPD) method and loaded by bellows, are used in the investigations. Continuous crack monitoring allows the effects of changing chemistry environments (e.g. normal water chemistry (NWC) versus hydrogen water chemistry (HWC)) to be assessed directly, as well as enabling the contributions of loading on cracking response to be evaluated.

The geometry of the C(T) samples is shown in Fig. 11.2. The specimens, with width $W = 16$ mm and thickness $B = 5$ mm, all have 8 mm long machined chevron notches and 10% side grooves such that B_{eff} is 4.47 mm. After machining, the C(T) specimens are fatigue pre-cracked in air, with the final pre-cracking conditions being 12 MPa m$^{1/2}$, $R = 0.1$, 22 Hz.

After installation in the test rig, the leads for crack propagation monitoring with the DCPD technique are spot welded to the specimen 'arm' extensions. The specimens are instrumented with two pairs of potential leads (i.e. four 1 mm mineral insulated (MI) cables) and one pair of current leads. All of the MI cables are single-wire Ni conductors with Inconel 600 sheath material. When spot-welded to the C(T) arm extensions with a specially designed cable connector, the MI Cable conductor and sheath are shorted together. Since the MI cable sheath will create current paths in parallel with the C(T) specimens due to random contacts between the sheaths at

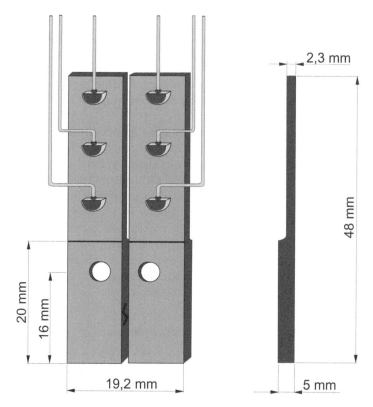

2,3 mm

48 mm

20 mm

16 mm

19,2 mm

5 mm

11.2 Drawing of C(T) sample for on-line crack growth rate measurements

various locations in the rig, the cables are firmly fixed to plates positioned 60 mm from the specimens to have a stable current leakage. Routing of the cables is then continued up through the rig to the pressure boundary/top seal assembly (which has a compressed graphite seal) and on to permanent signal cable junctions in the reactor hall.

Dynamic load is applied to the C(T) specimens by individually calibrated loading units which are equipped with bellows that are pressurised with helium gas through an outer system (Fig. 11.3). During irradiation, the specimens are subjected either to constant load or cyclic loading conditions. The cyclic loading with R (K_{min}/K_{max}) = 0.5, 0.6 or 0.7 is typically implemented once or three times in 24 h. Typically, the duration of an unloading-reloading cycle is ~900 s.

The arrangement of the C(T) specimens and other instrumentation in a typical test rig is shown in Fig. 11.4. Four specimens are positioned in the high fast neutron flux (booster fuel rod) region while a further sample is located in the upper test section in a low flux region. Neutron detectors at two elevations in the high flux region enable the fast neutron flux (and fluence accumulation) in the specimens to be computed. For measuring the ECP of the stainless steel pressure flask, Fe/Fe_3O_4, Pt and Pd electrodes are installed in the upper test section. Coolant thermocouples are placed at the inlet, middle and outlet of the test rig.

Figures 11.5 and 11.6 show results from a recently completed experiment conducted under BWR conditions, in which the effect of changing from oxidising

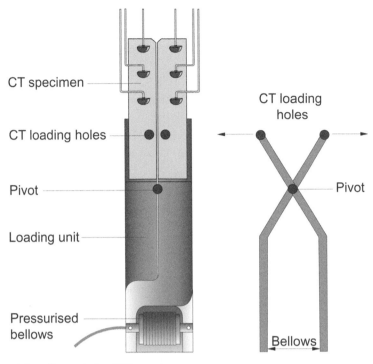

11.3 Loading unit for C(T) specimens

(5 ppm O_2) to reducing (2 ppm H_2) water chemistry was studied. For a high fluence ($2.5 \cdot 10^{22}$ n/cm^2) 304L stainless steel sample, there was no reduction in crack growth rate caused by the chemistry change (Fig. 11.5), whereas for a lower fluence ($6 \cdot 10^{20}$ n/cm^2) 347 stainless steel sample, a clear reduction in crack growth rate was seen (Fig. 11.6).

11.4   Crack initiation measurements

A crack initiation (integrated time-to-failure) study is currently being conducted in a BWR loop system. The number of specimen failures that occur in irradiated tensile stainless steel specimens as a function of the water chemistry (NWC versus HWC) is being evaluated. The aim is to provide information on the effectiveness of hydrogen additions in reducing the susceptibility to the initiation of cracks in high dose material [5,6].

A total of 60 irradiated miniature tensile test specimens, 30 to be exposed to NWC conditions and 30 to be exposed to HWC conditions, were prepared for the investigation. The specimens, which have a total length of 20 mm and a cylindrical (1 mm diameter, 4 mm long) gauge section, were all prepared from a $8 \cdot 10^{21}$ n/cm^2 304L SS control rod material from the Barsebäck 1 BWR. The fluence of this material is close to the maximum end of life fluence of BWR components such as the top guide and, because of the high dose, is more likely to exhibit susceptibility to stress corrosion cracking. In addition, the material will have reached saturation in terms of mechanical and microstructural property changes and is expected to behave in a more

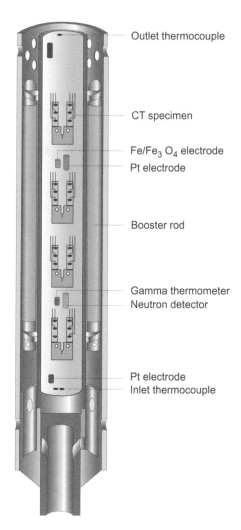

Outlet thermocouple

CT specimen

Fe/Fe$_3$O$_4$ electrode
Pt electrode

Booster rod

Gamma thermometer
Neutron detector

Pt electrode
Inlet thermocouple

11.4 Test rig for IASCC studies

consistent manner than lower dose materials where saturation has not yet been reached.

Thirty of the specimens are currently being exposed to the NWC conditions. The specimens are arranged in pairs at five elevations in the test assembly, with three pairs (i.e. six specimens) per elevation. Eighteen of the units are exposed to a high fast neutron flux and the remaining 12 are installed in a low flux position.

Each pair of specimens is equipped with a linear voltage differential transformer (LVDT) that enables on-line detection of failure (Fig. 11.7), and the specimens within each pair are identified by spacers placed internally in the bellows. In the event of specimen failure, the bellows collapse, with the extent of movement being recorded by the LVDT, which enables identification, on-line, of the failed sample.

During irradiation, constant load (corresponding to 76 to 97% of the irradiated yield strength (718 MPa) of the material) has been applied to the specimens by the bellows that are compressed by the system pressure.

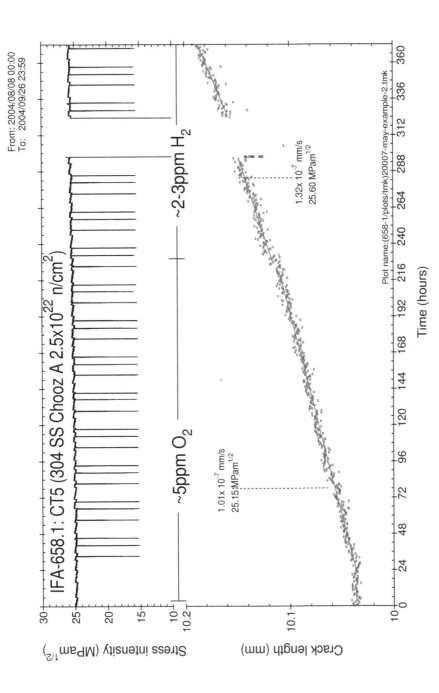

11.5 Crack growth rate measurements on a high fluence 304L stainless steel C(T) specimen. No effect of changing from oxidising to reducing water is observed

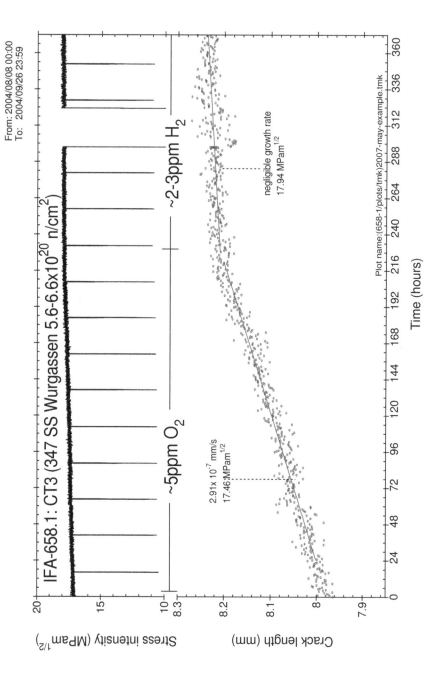

11.6 Crack growth rate measurements on a low fluence 347 stainless steel C(T) specimen. A clear effect of changing from oxidising to reducing water is observed

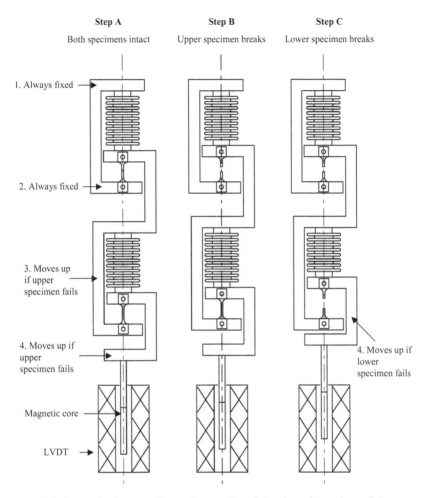

11.7 Schematic diagram illustrating on-line detection of specimen failure

An example of on-line detection of a specimen failure, from the change in the LVDT signal, is shown in Fig. 11.8, while Fig. 11.9 shows a photograph of the failed specimen after removal from the test rig.

11.5   Irradiation creep and stress relaxation measurements

Many bolts in reactor internals are stressed to high initial cold pre-loads, which when subjected to high operating temperatures over time, may result in loosening of the bolts and the pre-load can be lost. Radiation may also cause stress relaxation in highly stressed components.

The effects of irradiation on the creep/stress relaxation in austenitic stainless steels, commonly employed in commercial PWR reactors, are being evaluated [6]. The test materials are small tensile specimens (2.5 mm diameter and a gauge length of ~50 mm) prepared from unirradiated, solution annealed 304 and cold worked 316 stainless steels. The specimens are subject to constant displacement conditions during irradiation in an inert environment to fluences of ~0.25, 1 and $1.4 \cdot 10^{21}$ n/cm^2 (> 1 MeV).

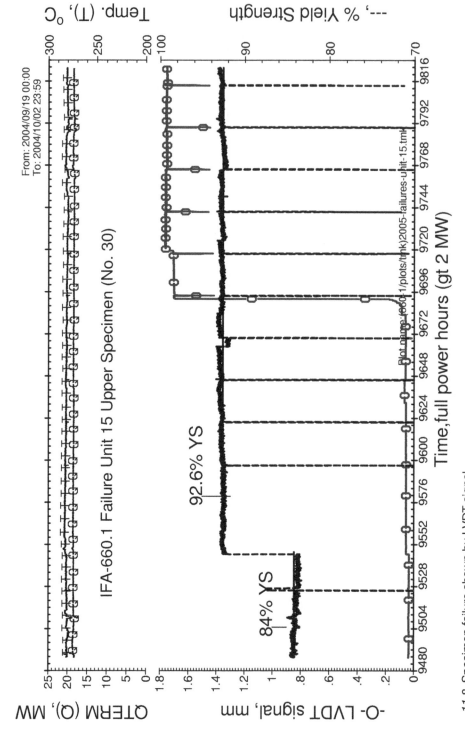

11.8 Specimen failure shown by LVDT signal

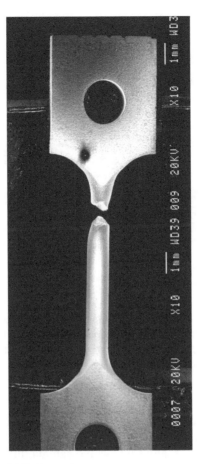

11.9 Photograph of failed specimen

Load (stress) is applied to the specimens via bellows that are compressed by gas pressure that is introduced into the chamber housing the bellows (Fig. 11.10). Constant displacement of the tensile specimens is maintained by monitoring sample elongation by LVDTs and adjusting (reducing) the applied load (stress) on the specimens on-line, by decreasing the pressure in the bellows housing units. In addition to the bellows gas lines, the test units are also equipped with gas lines that enable the specimen temperature to be varied in the range from 240 to 400°C, by altering the composition of the helium–argon gas mixture surrounding the specimens.

An example of the load reductions made on a 316N stainless steel lot specimen when length changes on the order of 2 µm are detected by the LVDT is shown in Fig. 11.11.

11.6   In-core ECP measurements

The electrochemical corrosion potential (ECP) of a corroding metal is the potential difference between it and the standard hydrogen electrode (SHE). ECP is determined by measuring the potential between the sample and a reference electrode, and adding the (calculated) potential versus SHE of the reference electrode. A reference electrode

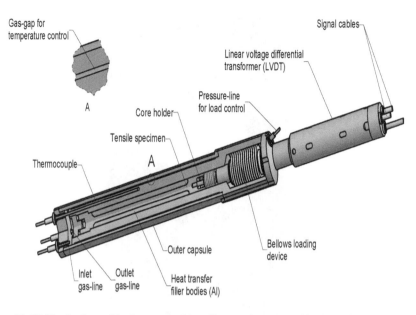

11.10 Illustration of instrumented tensile specimen used in the stress relaxation study. Instrumentation includes bellows that allows on-line variation of applied load (stress), temperature control through gas lines and monitoring of specimen elongation by an LVDT

is a half-cell that produces a stable and reproducible potential. For in-core measurements of ECP, reference electrodes must be capable of withstanding the high temperatures and pressures, and neutron and gamma fluxes within the reactor core.

The ECP is one of the most important measures of a corrosive environment, and is determined by a combination of the surface conditions of the specimen and concentrations of dissolved oxidants. The ECP is a key measurement when the performance of reactor structural materials is to be assessed or measures taken to optimise their integrity – many BWRs now inject hydrogen to maintain the ECP of in-core stainless steel components below the 'IASCC threshold' of -230 mV$_{SHE}$.

Perhaps the most critical aspect of the design of a reference electrode is the seal between the potential sensing element and the signal cable, which must be water-resistant. Temperature changes will cause mechanical stress on the electrode seals, signal cables, etc, and it can be assumed that the electrode lifetime will be dependent on the number of reactor power or loop operations causing large temperature changes in the loop coolant. In the electrodes designed and produced in Halden, the seal is constructed from two metal (Inconel) sub-units connected by a ceramic tube and has main dimensions of length 80–100 mm and diameter 12 mm. No ceramic-to-metal brazing is required [7].

To obtain reliable ECP measurements, it is recommended that two different types of reference electrode be used. Palladium (Pd) and iron/iron oxide (Fe/Fe$_3$O$_4$) electrodes are being developed for use under oxidising conditions; in addition, Pt electrodes can be used under reducing conditions.

The Halden platinum (Pt) electrode [8] is shown in Fig. 11.12. The potential-sensing element consists of a Pt cylinder and end plate. The large surface area of

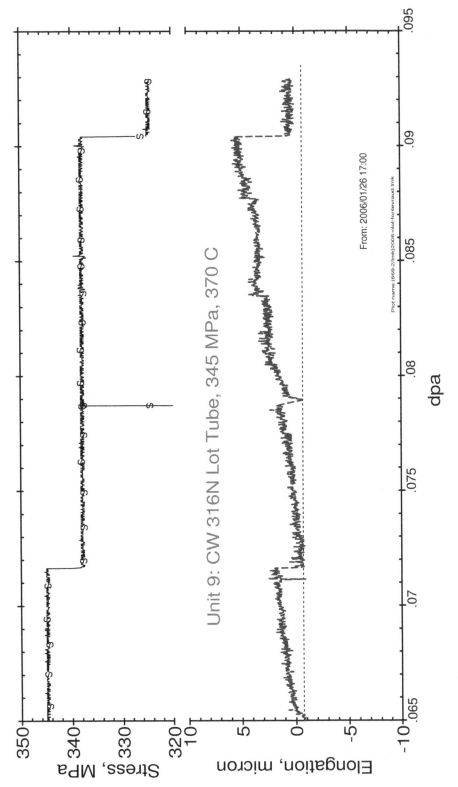

11.11 Example of on-line changes in stress that are made as specimen elongation on the order 2–6 μm is recorded by an LVDT

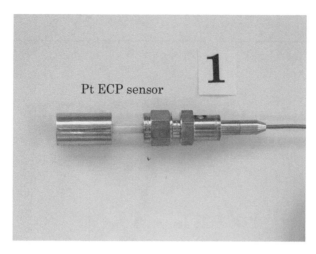

11.12 Halden platinum reference electrode

the platinum serves to reduce the effect of any mixed potentials caused by the other metal components.

Figure 11.13 shows an example of ECP data measured with a Pt electrode under PWR water chemistry conditions (3 ppm LiOH, 1000 ppm B as boric acid (H_3BO_4), 2 to 4 ppm H_2 and conductivity 25 to 30 μS/cm). The electrode gave reliable signals over the duration of the test, 340 full power days. This period included four reactor cycles, with several heating and cooldown periods within each, and shows that the seal arrangement is satisfactory. The measured potential between the stainless steel pressure flask (used as the sample) and the reference electrode was, as would be expected under reducing conditions, of the order of a few millivolts, i.e. both metals were at similar potentials. Under such conditions, the ECP of stainless steel primarily reflects the thermodynamic potential of the H_2/H_2O reaction, and values in the range from –770 to –830 mV_{SHE} were obtained.

The Halden/VTT palladium reference electrode [8,9] is shown in Fig. 11.14. The potential-determining reaction of the palladium hydrogen electrode is similar to that of the platinum hydrogen electrode or SHE [10]. In the SHE, hydrogen is introduced from an external source and the electrode potential is determined by the H_2 partial pressure and the pH of the electrolyte. Palladium differs from platinum in that it can hold hydrogen within its lattice; in aqueous solutions, hydrogen can be produced by cathodic charging of the Pd. When sufficient hydrogen is loaded, some will diffuse to the surface of the Pd, which can thus function as a hydrogen electrode. Hence, the advantage of the Pd electrode over the Pt electrode is that it can be used in oxygenated water. The Pd electrode is suitable for in-core measurements since the potential-determining species (H_2/H^+) are stable over the temperature range of interest and do not contaminate the environment (as would happen, for example, if an Ag/AgCl electrode were to fail).

The Pd electrode must be calibrated against another type of electrode to determine the surface coverage of palladium with hydrogen atoms. The surface coverage is analogous to the H_2 partial pressure used in the calculation of the potential of platinum electrodes, and thus allows calculation of the potential of the electrode.

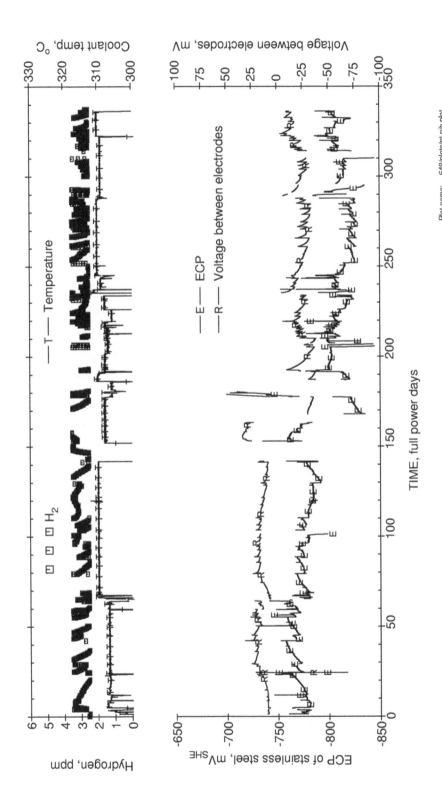

11.13 ECP of stainless steel under PWR conditions (3 ppm LiOH, 1000 ppm B), measured with a Pt reference electrode

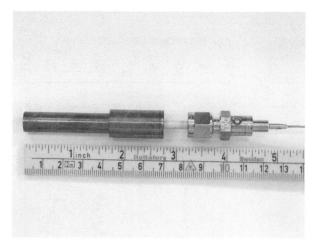

11.14 Halden/VTT palladium reference electrode

Initial in-core results [11] showed that the Pd electrode can produce a stable reference potential in oxygenated water (5 ppm O_2) (Fig. 11.15), and that it can measure changes in ECP due to changes in oxidant concentration. ECP measurements with the Pd electrode have also been performed in hydrogenated water; Fig. 11.16 shows the effect of coolant temperature on the surface coverage of hydrogen atoms on palladium [12].

Metal/metal oxide electrodes have been used for many years, and development of Fe/Fe_3O_4 electrodes has commenced at Halden. Whereas many designs use a ceramic tube made from yttrium partially stabilised zirconia (Y PSZ), the Halden studies have focused on magnesium partially stabilised zirconia (M PSZ) due to its higher strength, which should make it easier to construct a reliable seal. Tests have shown that M PSZ works as an oxygen conducting membrane and that the internal resistance of Mg-PSZ (Fe/Fe_3O_4 electrode) is low enough to allow in-pile measurements. A prototype electrode is shown in Fig. 11.17.

11.7   In-core conductivity measurements

The inter-relationship between ECP and conductivity, and hence between conductivity and crack propagation, is well established. Most on-line techniques for monitoring water quality are designed for use at room temperature. However, room temperature measurements may not be representative of the conductivity at elevated temperature. Although high-temperature conductivity can be modelled, the calculations require as input the high-temperature concentrations of dissolved ions, which may not be available. Hence, the advantages in measuring conductivity at the temperature of interest, for both precise determination of water quality and to study the effect of impurities on corrosion, are clear.

A novel conductivity electrode (Fig. 11.18) has been developed that can be installed in in-core pressure flasks and hence measure conductivity at operating temperature and pressure, and in the presence of water radiolysis products. The sensor is a modification of the Halden platinum ECP electrode, in which the Pt sensing element is surrounded by a second Pt cylinder. An additional signal cable is fitted through

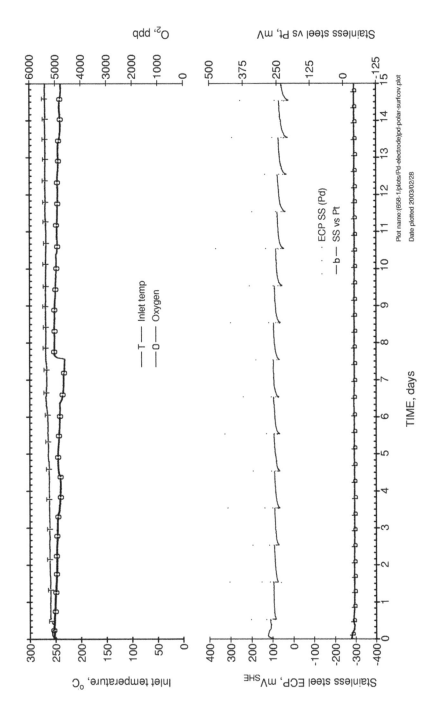

11.15 ECP of stainless steel, measured with a Pd electrode, during steady-state operation

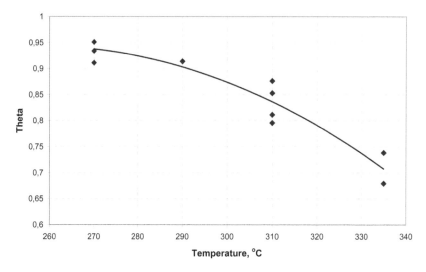

11.16 Surface coverage of hydrogen on palladium versus temperature for hydrogenated water chemistry (250 ppb LiOH, 5 ppm H_2)

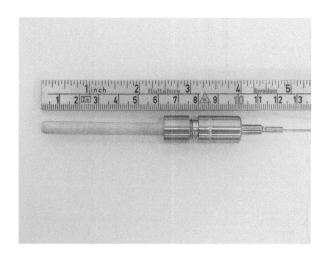

11.17 Prototype Fe/Fe_3O_4 reference electrode

which a current is passed, and the conductivity of the coolant is determined from the measured voltage difference between the two Pt cylinders. Use of the Halden leak-tight mechanical seal enables the monitoring of conductivity in-core. It is envisaged that the sensor will give the most useful data under BWR conditions, since impurity leakages are more readily monitored at high temperature, due to increased ionic mobilities. Under PWR conditions, conductivity is dominated by LiOH and boric acid, and the effects of impurities may not be distinguishable.

The prototype cell functioned reliably in a test conducted under PWR conditions (2 ppm LiOH, 1200 ppm B, 2.5 to 3 ppm H_2) for a period of 45 days (Fig. 11.19). A solution conductivity of approximately 200 µS/cm was measured at 335°C; the

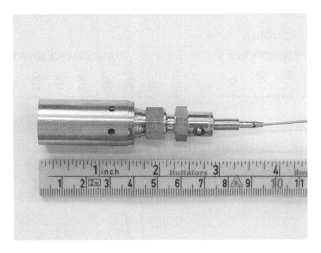

11.18 In-core conductivity cell

corresponding low temperature conductivity, measured in the loop sampling system, was approximately 20 µS/cm. The measured conductivity agreed well with literature data of the high-temperature conductivity of LiOH solutions.

The sensor may give more useful data under BWR conditions, where water radiolysis products will have a larger effect on the in-core conductivity. Chemical additions to the coolant are planned in several forthcoming BWR IASCC studies; conductivity measurements in these experiments will be conducted to investigate whether the cell gives useful results in water with a lower conductivity than in PWR coolant.

11.8   Techniques under development to measure fuel cladding corrosion on-line

Traditionally, corrosion of fuel rod cladding in reactor cores has been studied by taking measurements of oxide thickness or sample weight during reactor shutdowns. The main limitation of this method is that only average corrosion rates can be determined. The use of on-line measurements would deliver more information on corrosion processes and would assist in the determination of corrosion mechanisms.

11.8.1   On-line potential drop corrosion monitor

The DCPD method, which has been developed and used to measure crack growth on-line in the Halden IASCC programme, has been adapted for on-line measurement of cladding corrosion. Current and potential wires are attached to the end plugs of a fuel rod, and potential drops caused by the passage of an applied current are measured at different positions on the end plugs (Fig. 11.20). The potential drops are related to cladding thickness through out-of-pile calibration. With this technique, cladding thickness can be measured to an accuracy of ±2 µm.

Qualification and testing of the method under in-core conditions will be conducted. Oxide thicknesses will be compared with measurements performed using standard techniques such as eddy current and SEM.

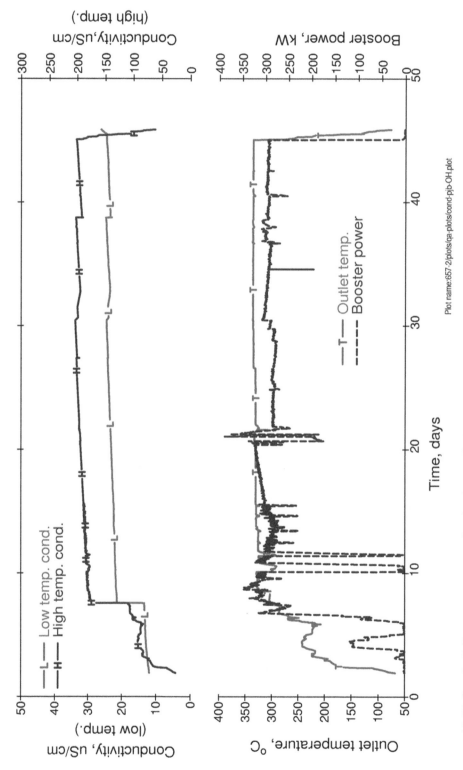

11.19 Conductivity measurements, showing in-core (high temperature) and out-of-core (low temperature) data

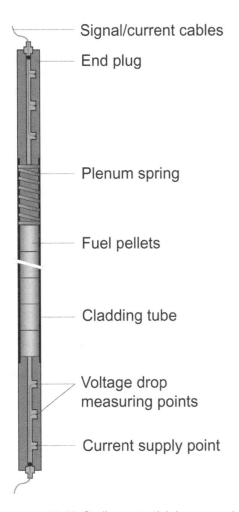

Signal/current cables

End plug

Plenum spring

Fuel pellets

Cladding tube

Voltage drop
measuring points

Current supply point

11.20 On-line potential drop corrosion monitor

11.8.2 Electrochemical impedance spectroscopy

In electrochemical impedance spectroscopy (EIS) measurements, a variable (a.c.) voltage is applied to the sample and its impedance (ratio of voltage to current) is measured as the frequency of the applied voltage changes. The technique possesses several advantages. Methods using a.c. use low perturbation signals that do not disturb the system under test, as do d.c. methods. The technique can be used in low conductivity media, such as BWR type coolant, and is non-destructive.

Preliminary development work, in conjunction with the University of Gothenburg, Sweden, was conducted to assess the feasibility of monitoring oxide growth on Zircaloy fuel cladding on-line. Measurements were taken using cable lengths of up to 20 m to study any adverse effects caused by the large separation of the sample and measuring equipment. Oxide thicknesses were calculated from an equivalent electrical circuit, in which the insulating ZrO_2 is modelled by a capacitor with plate separation equal to the oxide thickness.

Initial in-core measurements will be performed under PWR conditions, i.e. in water with a relatively high conductivity. The sample (or working electrode) will consist of either a coupon or a tube placed inside a cylindrical platinum mesh counter electrode. Platinum reference electrodes will also be installed. The measurements will be conducted using commercial computer software loaded onto computers located in the reactor hall.

Provisionally, it is proposed to test Zircaloy-4 and Zircaloy-2. The former will allow comparison of the corrosion rate data with corrosion models, while use of the two materials will allow demonstration of any effects of material composition on corrosion behaviour. Other alloys, either those used currently in LWRs or under development, can also be tested. Further experiments will be conducted under BWR water chemistry conditions.

References

1. W. Wiesenack, 'OECD Halden Reactor Project fuel testing capabilities and high burn-up data for modelling and safety analysis', IAEA research coordination meeting on fuel modelling at extended burnup (FUMEX) (Bombay, India, 1–5 April 1996).
2. T. M. Karlsen and E. Hauso, 'Qualification and application of instrumented specimens for in-core studies on cracking behaviour of austenitic stainless steels', in *Proc. 9th Int. Conf. on Environmental Degradation of Materials in Nuclear Power Systems – Water Reactors* (Newport Beach, CA, USA, August 1999). NACE, Houston, TX, USA.
3. T. M. Karlsen, M. McGrath and E. Kolstad, 'Halden research on Zircaloy cladding corrosion', in *Water Chemistry and Corrosion Control of Cladding and Primary Circuit Components*, in Proceedings of a Technical Committee meeting (Vltavou, Czech Republic, 28 September–2 October 1998), IAEA-TECDOC-1128. IAEA.
4. P. J. Bennett, E. Hauso, N.-W. Høgberg, T. M. Karlsen and M. A. McGrath, 'In-core materials testing under LWR conditions in the Halden reactor', in *Proc. Chimie 2002* (Avignon, France, 22–26 April 2002). SFEN, 2002.
5. T. M. Karlsen, P. J. Bennett and N.-W. Høgberg, 'In-core crack growth rate studies on irradiated austenitic stainless steels in BWR and PWR conditions in the Halden reactor', in *12th Int. Conf. on Environmental Degradation of Materials in Nuclear Power Systems – Water Reactors*. NACE, Houston, TX, USA, 2005.
6. T. M. Karlsen, N.-W. Høgberg and R. van Nieuwenhove, Results from irradiation assisted stress corrosion cracking and stress relaxation studies performed in the Halden reactor, Fontevraud 6 (18–22 September 2006). SFEN.
7. B. Beverskog, L. Lie, N.-W. Høgberg and K. Mäkelä, 'Verification of the miniaturised in-core Pd reference electrode operation in out-of-core and in-core location at Halden', in *Proc. Water Chemistry of Nuclear Reactor Systems 8* (Bournemouth, UK, 22–26 October 2000). BNES, 2000.
8. P. J. Bennett, B. Beverskog, N.-W. Høgberg, T. M. Karlsen and H. Thoresen, 'In-core ECP measurements under LWR conditions in the HBWR', in *Tenth Int. Conf. on Environmental Degradation of Materials in Nuclear Power Systems – Water Reactors*. NACE, Houston, TX, USA, 2001.
9. P. J. Bennett, M. McGrath, K. Bagli and M. Dymarski, 'Measurements of carbon steel ECP and critical deuterium concentration under CANDU conditions in the Halden Reactor', in *12th Int. Conf. on Environmental Degradation of Materials in Nuclear Power Systems – Water Reactors*. NACE, Houston, TX, USA, 2005.
10. R. W. Bosch, R. van Nieuwenhove, W. F. Bogaerts and F. Moons, 'Development of high temperature, high pressure reference electrodes for in-pile use', in EUROCORR '98 (28 Utrecht, Netherlands, September–1 October 1998). EFC, 1998.

11. P. J. Bennett, 'In-core ECP measurements in oxygenated water in the Halden Reactor using a palladium reference electrode', in *Int. Conf. on Water Chemistry of Nuclear Reactor Systems* (San Francisco, CA, 11– 4 October 2004). EPRI.
12. P. J. Bennett, In-core ECP measurements under PWR conditions using a palladium reference electrode, Fontevraud 6 (18–22 September 2006). SFEN.

12

Corrosion monitoring applications in Taiwan's nuclear power plants

Hsuan-Chin Lai and Wei-Yun Mao

Material and Chemical Research Lab., Industrial Technology Research Institute, 195 Chung Hsing Rd., Sec.4 Chu Tung, Hsin Chu, Taiwan (R.O.C.)

HCLai@itri.org.tw; WYMao@itri.org.tw

Charles Fang Chu

Dept. of Nuclear Generation, Taiwan Power Company, No.242, Sec.3, Roosevelt Rd., Zhongzheng District, Taipei City 100, Taiwan (R.O.C.)

12.1   Introduction

On-line corrosion monitoring methods have been applied extensively for system condition diagnosis, material degradation resistance evaluations, and water chemistry optimisation in Taiwan's boiling water reactor (BWR) and pressurised water reactor (PWR) nuclear power plants. Electrochemical corrosion potentials (ECP) in Chinshan and Kuosheng BWR water circulation loops are measured to evaluate the extent of ECP protection provided by hydrogen water chemistry (HWC). On-line crack advance surveillance equipment can continuously monitor the effectiveness of HWC in mitigating stress corrosion cracking (SCC) of stainless steel components. In the Maanshan PWR, ECP measurements in secondary systems are used both as a sensitive diagnostic tool for system oxygen abnormalities and as a reference for hydrazine concentration control. On-line monitors are installed for service water loops to detect the corrosion behaviour of materials in the system under different water chemistry conditions. In decontamination processes, on-line corrosion monitors are also implemented to ensure the integrity of systems and components. In addition, a pipe-thinning monitor is installed in Maanshan plant to continuously monitor the thickness reduction of carbon steel (CS) components during normal plant operation. On-line corrosion monitors have proven to be very useful tools within the nuclear power monitoring systems currently installed and have fulfilled their specific purpose in the operation of Taiwan's power plants. This chapter will provide a general description of on-line corrosion monitoring systems developed and currently installed in Taiwan and their mode of operation.

12.2   Application examples

12.2.1   ECP monitoring in BWR

HWC is implemented in BWR systems to mitigate intergranular stress corrosion cracking (IGSCC) problems of stainless steel components in the reactor pressure vessel and recirculation piping systems. The effectiveness of HWC in BWR is mainly evaluated through the measurement of the ECP of stainless steel components. The general consensus is that an ECP below $-230\ \text{mV}_{SHE}$ is required to provide sufficient protection against IGSCC. In order to determine an adequate hydrogen injection level, HWC verification checks have to be carried out in both Chinshan and Kuosheng plants. Two such tests are being performed at each plant. One uses the ECP package installed by the commercial vendor at both the reactor pressure vessel (RPV) bottom drain line and reactor recirculation system (RRS) flange locations to confirm the effect of hydrogen injection. The additional test for hydrogen effectiveness is implemented by MCL/ITRI to measure the ECP in a side-stream loop. Figure 12.1 shows the ECP monitoring systems installed in both plants. Ag/AgCl reference electrodes are used to measure the SS304 ECP.

The location for ECP measurement in Chinshan unit 1 is in the side-stream loop of the RPV bottom head drain line. Figure 12.2a shows the schedule of verification tests and the results of ECP measurements at this location. Under normal water chemistry (NWC) conditions, the ECP measured in the side-stream loop is about $100\ \text{mV}_{SHE}$.

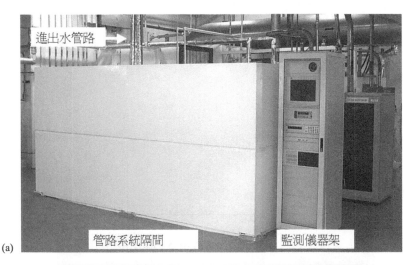

(a)

(b)

12.1. ECP measurement systems in (a) Chinshan Unit 1 and (b) Kuosheng Unit 2

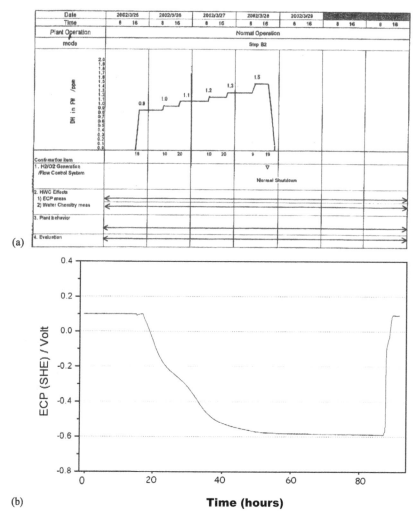

12.2. Hydrogen injection at Chinshan Unit 1, (a) the HWC verification test schedule, and (b) ECP measurement result

Under HWC conditions, the ECP decreases with increasing dissolved hydrogen in the feed water and approaches a stable value (Fig. 12.2b). To make the ECP value drop below –230 mV$_{SHE}$, the hydrogen injection should be adjusted to a level of more than 1.0 ppm dissolved hydrogen in the feed water.

The location of ECP measurement in Kuosheng unit 2 is in the side-stream loop of the reactor water clean up pump outlet. The water at this location is a mixture from the recirculation system and the RPV bottom head drain line. Figure 12.3 shows the schedule of verification tests and the resulting ECP measurements. Under NWC conditions, the ECP measured in the side-stream loop is about 40 mV$_{SHE}$. Upon hydrogen injection, the ECP decreases with increasing hydrogen level and gradually approaches a stable value (Fig. 12.3b). However, the ECP is still higher than –230 mV$_{SHE}$ while dissolved hydrogen in the feed water is lower than 0.9 ppb. To reach the desired ECP value below –230 mV$_{SHE}$, more than 0.9 ppm dissolved hydrogen is needed in the feed water.

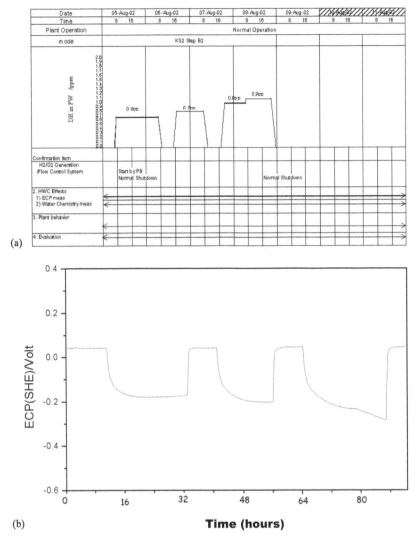

12.3. Hydrogen injection at Kuosheng Unit 2, (a) the HWC verification test schedule, and (b) ECP measurement result

12.2.2 Crack growth rate measurements

As discussed above, the effectiveness of hydrogen injection is indicated by the ECP value. However, the benefit of HWC should manifest itself by the prevention and reduction of SCC in stainless steel components. MCL/ITRI has installed a bellows-driven tensile machine for measuring the crack growth rate (CGR) of Type 304 stainless steel in reactor coolant at Chinshan nuclear power plant. The effects of varying stress intensity factors, water chemistry parameters, degree of sensitisation, and different surface conditioning on the CGR can all be evaluated in-situ. The CGR test data of as-received type 304 stainless steel in the Chinshan reactor water under NWC conditions is shown in Fig. 12.4. The crack in the as-received stainless steel specimen propagated at a rate of about $1.2 \cdot 10^{-8}$ mm/s at 25 MPa m$^{1/2}$ under NWC, as

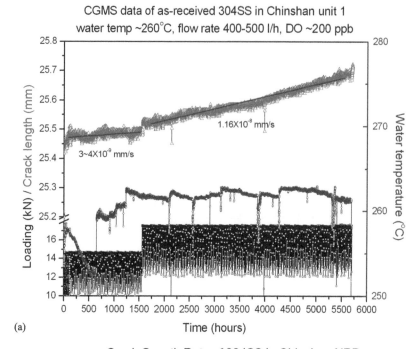

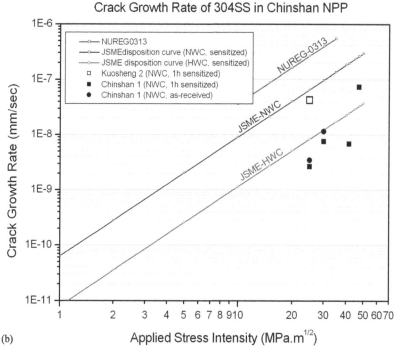

12.4. In-plant crack growth rate monitoring for (a) as-received type 304 stainless steel compact tension specimen under NWC conditions at Chinshan Unit 1, and (b) CGR data of as-received and sensitised type 304 stainless steel at varied applied stress intensities under NWC depicted in Fig. 12.4a. The CGR test data for as-received and sensitised stainless steel in Chinshan unit 1 reactor coolant showed that the cracks grew at a rate well below the JSME disposition curve [1], as shown in Fig. 12.4b. The corresponding data from the Kuosheng plant was close to the available data in the literature [2]. Taipower injects 0.5 ppm of hydrogen into the reactor water during the first phase of HWC. The CGR of stainless steel is lowered by the injected hydrogen, as expected, but the full effectiveness on SCC mitigation will be confirmed quantitatively after HWC implementation. The CGR measurements will also be conducted using MCL's in-laboratory facility in controlled environments simulating NWC and HWC. The field data can be compared with laboratory data.

12.2.3 Electrochemical noise monitoring during a decontamination process

The piping of the reactor recirculation system (RRS) and the reactor water clean up system of Taiwan Power Company's Kuosheng nuclear power plant were decontaminated using the CORD process developed by Framatome ANP GmbH during the October 2001 outage. To verify the adequacy of process control, and to detect any possible corrosion damage, the material corrosion behaviour was continuously monitored during all stages of the chemical decontamination. Three kinds of specimen were adopted for this corrosion monitoring, including corrosion coupons for weight-loss measurements, electrochemical noise (EN) specimens for on-line corrosion monitoring, and wedge-opening-loaded (WOL) specimens for stress corrosion evaluation. The test materials included Type 304 stainless steel (SS) with both sensitised and as-received thermal history, Type 308 weld filler, Type CF8 cast SS, nickel-base Alloy 182 weld filler, Inconel 600, Stellite 6 hard-facing alloy, NOREM low-cobalt hard-facing alloy, and A106B CS. All coupons were pre-filmed in an NWC water environment with 1 ppm dissolved oxygen for 72 h at 288°C, except one coupon of 304Sen (i.e. sensitised type 304 SS) which was pre-filmed in simulated HWC water. The EN probe for on-line corrosion monitoring is a three-electrode design, including working electrode (WE) of a circular specimen, counter electrode of a C-shaped specimen, and reference electrode of Pt wire. Two 304 SS working electrodes (WEs) were used in the RRS-A piping for CORD UV process evaluation, and two A106B CS WEs in the reactor water clean up system for CORD CS. To simulate the crevice condition in the real pipe, the ceramic washer was tightened onto the surface of the 304 SS WE as illustrated in Fig. 12.5.

The EN measurement used the ACM AutoZRA measurement system which provides a three-electrode configuration, i.e. WE1, WE2 and reference electrode (RE). The current noise was measured between WE1 and WE2. The voltage noise was measured between the RE and the galvanic potential of WE1 and WE2. The corrosion rate of the EN specimen was calculated by noise resistance, which is the ratio of standard deviation of voltage noise (VSD) over current noise (ISD). The physical meaning of noise resistance is similar to polarisation resistance. Therefore, the following equations are adequate for the calculation of its corrosion rate.

$$I_{corr} = \frac{\beta_a * \beta_c}{2.3 * R_p * (\beta_a + \beta_c)}$$

$$CR = \frac{0.13 * I_{corr} * EW}{A * d}$$

12.5. (a) EN specimen; (b) electrochemical specimen for crevice corrosion evaluation

where I_{corr} = corrosion current (μA), β_a = anodic Tafel constant, β_c = cathodic Tafel constant, R_p = polarisation resistance, CR = corrosion rate (mpy), EW = equivalent weight, A = surface area of specimen (cm^2), d = density of specimen (g/cm^3).

The EN current and voltage signals clearly characterised three decontamination cycles of CORD UV and the decontamination steps in each cycle. Figure 12.6 shows the current and potential data of 304Sen (NWC) during the second CORD UV process. The estimated metal corrosion rates by EN measurement in relation to the decontamination process are depicted in Fig. 12.7 for 304Sen (NWC and HWC) in the CORD UV process, and in Fig. 12.8 for galvanic CS-CS and CS-SS in the CORD CS process. The estimated metal loss was consistent with the weight loss from corrosion coupons. Comparing Fig. 12.7 with 12.9 and Fig. 12.8 with 12.10, the estimated metal loss by EN measurements agreed fully with the decontamination efficiency on the existing SS oxide films pre-formed in NWC and HWC and galvanic corrosion between SS and CS. Figure 12.7 also demonstrates that the corrosion of 304Sen (NWC) in the CORD UV process was higher than that of 304Sen (HWC). This result was further confirmed by the surface brightness in metallurgical examination of corrosion coupons. The extent of galvanic corrosion is clearly depicted in Fig. 12.8, which compared the corrosion of a galvanic couple of A106B CS-304 SS and uncoupled A106B CS during the CORD CS process. This result is consistent with the weight loss of corrosion coupons.

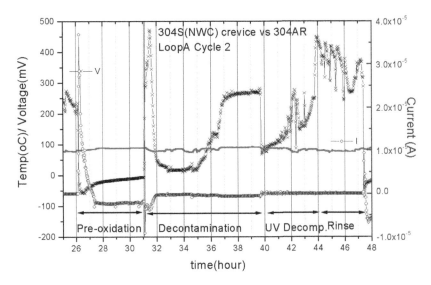

12.6. Signals of current and voltage from 304Sen (NWC) during the second CORD UV process

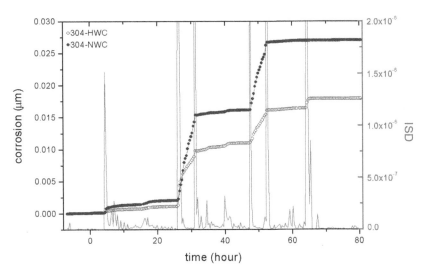

12.7. Corrosion attack on 304Sen (preconditioned in NWC) and 304Sen (preconditioned in HWC) during the CORD UV process

Most interestingly, the estimated trend of accumulated metal loss corresponds nicely to the total removed activities. Figure 12.9 plotted the accumulated removed activity in the RRS-A piping for the CORD UV process, and Fig. 12.10 recorded the accumulated removed activity in the reactor water clean up piping for the CORD CS process. Comparing Fig. 12.7 with Fig. 12.9 for the CORD UV process and Fig. 12.8 with Fig. 12.10 for the CORD CS process, the increase in removed activity and the accumulated metal loss followed a similar trend.

In summary, the measured metal losses from nine coupon materials did not indicate any unexpected or intolerably high corrosion damage from the CORD UV or

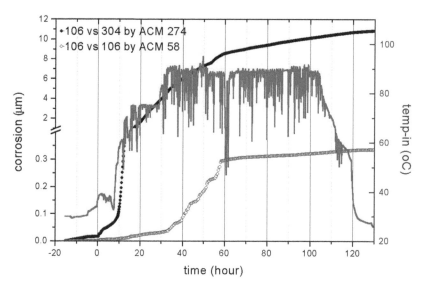

12.8. Comparison of change in corrosion attack and temperature of a galvanic couple of A106B CS-304 SS and uncoupled A106B CS during the CORD CS process

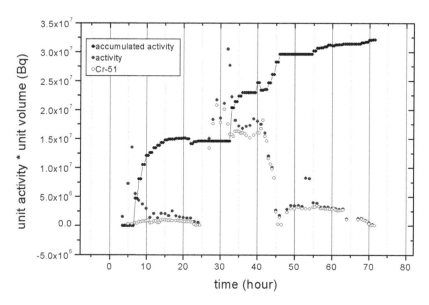

12.9. Plot of the accumulated removed activity, activity per unit volume, and Cr-51 activity in the RRS-A piping during the CORD UV process

CORD CS processes. The existing cracks did not show signs of propagation during the decontamination. The detection ability of metal corrosion by EN measurements using a three-electrode design was successfully demonstrated during the chemical decontamination process.

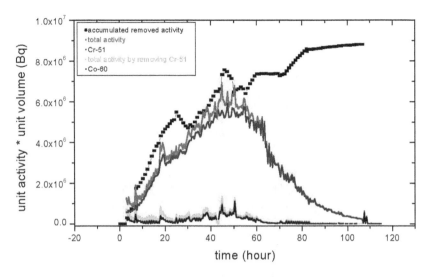

12.10. Accumulated removed activity, activity per unit volume, Cr-51 and Co-60 activity in the reactor water clean up piping during the CORD CS process

12.2.4   ECP monitoring in PWR secondary water

An electrochemical potential monitoring system is installed in the high-pressure heating region of the secondary feed water system (Fig. 12.11) in Maanshan PWR Unit 1 and 2. The installation is located where feed water enters into the steam generator, therefore closely reflecting the redox state of the feed water in contact with the steam generator tubing. Two Ag/AgCl reference electrodes with electrolytes of 0.01 N KCl and $5 \cdot 10^{-5}$ N KCl have been selected for ECP measurement. Three working electrodes of platinum, Inconel 600, and CS are used to monitor the secondary materials of concern. The platinum working electrode serves as a sensitive indicator for redox change in the secondary water chemistry. The electrochemical potentials of each WE are measured continuously along with the water chemistry data, such as hydrazine concentration, feed water oxygen content, condensate oxygen content, secondary water pH, etc.

Figure 12.12 shows the ECP and hydrazine concentration variations in Unit 1 secondary water. Clearly, ECP responds very quickly to slight variations of hydrazine concentration in the range between 15 and 40 ppb. In Unit 2, ECP responds similarly to hydrazine concentration (Fig. 12.13). As the condenser polisher is operating at full flow, the hydrazine concentration is increased to above 150 ppb. The ECPs of all three WEs immediately exhibit sudden downward shifts. As the hydrazine concentrations drift back to lower levels, the ECPs swiftly return to higher values. The ECP is accurately reflecting the actual redox state of the secondary water in a much more instantaneous and sensitive manner than the oxygen sensor itself. Thereby, the ECP monitor can be used as an oxygen abnormality or leakage sensor for the secondary system.

SCC of nickel-based alloys can be alleviated if the secondary water is changed to reducing conditions. And the erosion corrosion of CS feed water piping is substantially reduced when CS is maintained above -700 mV$_{SHE}$ [3]. An optimal ECP range can be controlled for materials in the secondary system by adjusting the hydrazine contents to minimise the corrosion-induced degradations.

(a)

(b)

12.11. ECP measuring system in PWR secondary feed water; (a) ECP data acquisition system; (b) ECP reference electrode and piping system

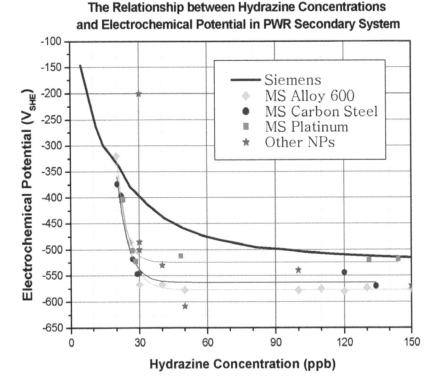

12.12. Relationship between ECP and hydrazine concentrations in Maanshan PWR Unit 1 secondary system

12.2.5   Closed cooling water system corrosion monitor

The continued operation and structural integrity of closed cooling water (CCW) systems are crucial for operating nuclear power plants. To date, environmentally compatible molybdate and nitrite corrosion inhibitors, replacing chromate inhibitor, have been used for mitigating the corrosion of structural materials in CCW systems. Four sets of on-line corrosion surveillance systems utilising the linear polarisation resistance (LPR) method were installed in the nuclear component closed cooling water system (NCCCW) and the turbine plant closed cooling water system (TPCCW) of Kuosheng nuclear power plant. Figure 12.14 shows the layout. A laboratory study was also conducted to obtain the optimal water chemistry condition in which lower corrosion rates of CS could be controlled. The effectiveness of the lowered corrosion inhibitor concentration programme in CCW systems is continuously monitored by the existing corrosion surveillance systems.

12.2.6   Pipe-thinning monitor

Wall thinning of CS piping caused by erosion corrosion is a safety problem in nuclear power plants. A pipe-thinning monitor has been developed to continuously monitor the thickness reduction of components during normal plant operation. This pipe-thinning monitor, which involves attaching metal wires onto the outer surface of a component to measure the reduction of thickness caused by internal corrosion,

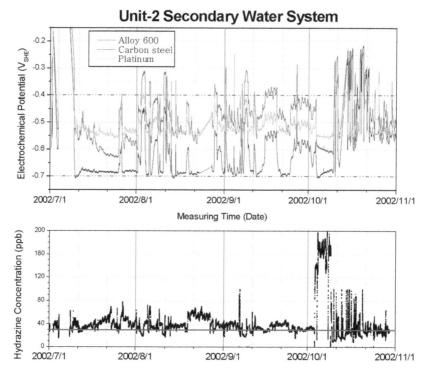

12.13. ECP measurement in Maanshan PWR Unit 2 secondary system

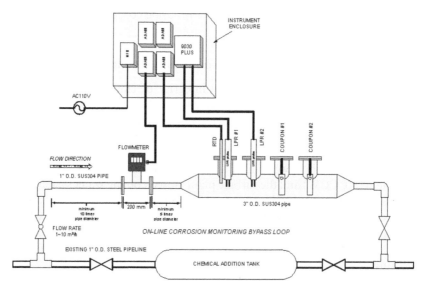

12.14. Layout of on-line LPR corrosion monitoring for component materials in Kuosheng's NCCCW and TPCCW systems

is schematically illustrated in Fig. 12.15. The potential drop technique is utilised to measure the loss of metal. Electric current flowing through the metal generates a potential drop due to ohmic resistance. Corrosion of the metal changes its geometry and, in turn, its electrical resistance. Consequently, the magnitude of the potential

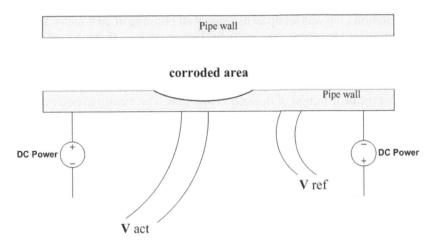

12.15. Schematic diagram for potential drop measurement

drop across measurement points changes with the size of the internally corroded area. Monitoring potential drop continuously can provide a measure of the degree of reduction of wall thickness.

Figure 12.16 shows the instrumentation. A current source supplies steady d.c. current to generate a potential drop across the sample. Typically, the electric current is adjusted to generate a potential drop higher than $50\,\mu V$. The potential drop is measured using a high precision voltmeter and then converted to a thickness index using calibrated software on a personal computer.

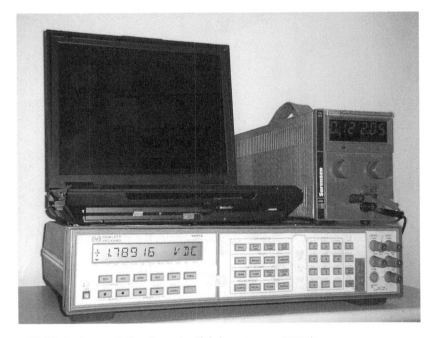

12.16. Instrumentation for potential drop measurement

A circulation loop was set up to measure wall thinning of an 8-inch CS elbow. The inner surface of the elbow was partially immersed in a flowing acid solution by adjusting the flow rate. Current leads and signal wires were spot-welded onto the pipe outer surface for potential drop measurement. The wiring for potential drop measurement was arranged on the outer surface of the elbow as shown in Fig. 12.17 to measure the distribution of thickness reduction. The measured potential drop was then converted to thickness using a calibration curve.

Figure 12.18 shows the interior of the elbow during water flow. The inner surface is partially covered with acid solution. Figure 12.19 demonstrates the percentage reduction at each measurement location on the elbow. There is no thickness reduction at locations A1, B1, C1 and C2. These locations are on the upper right side of the water flow as shown in Fig. 12.18. The acidic solution did not flow through the inner surface of these locations. On the other hand, apparent reduction occurs in regions close to locations A5, B5 and C5. Figure 12.18 shows that water flow directly washes over the inner surface of this area. The distribution of thickness reduction also agrees with ultrasonic testing inspection results.

In summary, this thickness-monitoring technique, which requires the attachment of metal wires onto the component outer surface to measure thickness changes caused by internal corrosion, shows high measurement sensitivity and excellent stability. Results of pipe tests confirm that this technique can effectively measure both the thickness reduction and the distribution of internal corrosion at the piping wall. This pipe-thinning monitor is also installed in Maanshan's condensate water line.

12.3   Conclusions

On-line corrosion monitoring systems are being put to many uses in Taiwan's nuclear power plants, ranging from water chemistry control to process monitoring and

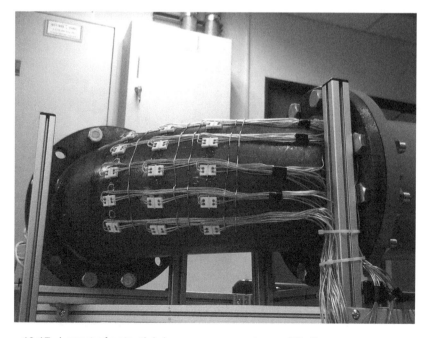

12.17. Layout of potential drop measurement on a CS elbow

12.18. Interior of the elbow during test

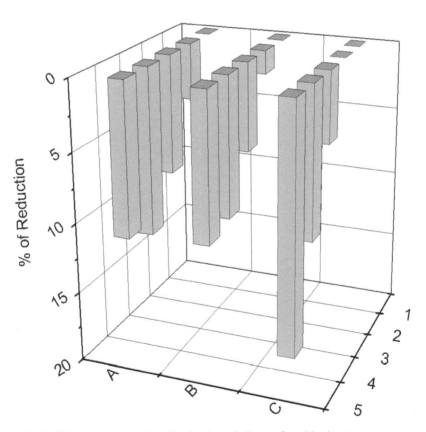

12.19. Thickness reduction distribution of elbow after thinning test

optimisation of material behaviour in BWR and PWR environments. More applications of crevice simulation in steam generator tubing, ECP monitoring in PWR primary water, and corrosion detection in service water loops are in the development and planning stages. Hopefully, with the aid of on-line continuous corrosion monitoring systems, material degradation due to electrochemical interactions with the environment can be kept to a minimum thus ensuring the structural integrity and functional properties of the nuclear components.

References

1. Maintenance Code of *Electricity Generating Nuclear Power Components*, JSME SNA 1-2000. JSME.
2. P. Andresen and L. M. Young, 'Characterization of the roles of electrochemistry, convection, and crack chemistry in stress corrosion cracking', in *Proc. 7th Int. Symp. on Environmental Degradation of Materials in Nuclear Power Systems – Water Reactor*. NACE, Houston, TX, USA, 1995, 579.
3. O. de Bouvier, M. Bouchacourt and K. Fruzzetti, 'Redox conditions effect on flow accelerated corrosion: Influence of hydrazine and oxygen', in *Chimie 2002, International Conference of Water Chemistry in Nuclear Reactors Systems*. SFEN, Avignon, paper 117.

A new technique for in-plant high mass transfer rate ECP monitoring

Anders Molander and Mats Ullberg

Studsvik Nuclear AB, S-61182 Nyköping, Sweden

anders.molander@studsvik.se

13.1   Introduction

Feed water chemistry in pressurised water reactors (PWRs) is important for mitigating corrosion and corrosion product transport. Recently, feed water chemistry has been given an increased focus and new approaches have been published. Many PWRs have experienced some chemistry-related corrosion concerns (such as danger of flow-accelerated corrosion, sludge burden, ingress of oxidants and reducible iron oxides into the steam generators), as well as some environmental concerns (in particular, hydrazine emissions during start-ups). In order to avoid capacity loss due to various corrosion effects, it is necessary to examine the possibilities of further improvements in secondary coolant chemistry. In particular, a slight increase in the feed water oxygen content is at present in focus to decrease flow-assisted corrosion and corrosion product transportation. It is well-known that the presence of small amounts of oxidants in high-temperature systems is difficult to detect and monitor. This paper describes electrochemical corrosion potential (ECP) measurements in PWR secondary systems and presents an improved monitoring technique to detect such low oxygen levels.

ECP measurements were introduced in nuclear power plants when intergranular stress corrosion cracking (IGSCC) in sensitised Type 304 stainless steel was the dominant type of failure in boiling water reactors (BWRs) [1]. The ECP was the prime parameter controlling IGSCC. Hydrogen water chemistry (HWC) was developed and introduced as a remedy. Hydrogen was added to the feed water to suppress the radiolysis and to decrease the ECP below the critical potential for IGSCC. The ECP monitoring equipment was first installed in remote sampling systems and the desired response to hydrogen was easily obtained. With time, the effect of hydrogen peroxide decomposition and other reactions in sampling lines was understood and monitoring locations with short transport time or in-situ measurements were implemented. Corrosion potential measurements developed into a routine technology necessary for safe reactor operation.

ECP measurements have also been performed in PWR systems and mainly the feed water system on the secondary side of PWRs [2–4]. The measurements performed so far have shown that electrochemical measurements are very sensitive tools to detect and follow oxygen transients in the feed water system. The redox conditions of the secondary system in a PWR are commonly assumed to be given by hydrazine addition and the residual oxygen concentration. Conventional water chemistry surveillance therefore includes monitoring of these species. However, the analyses are performed at ambient temperature and the sampling flow of feed water is often transported in fairly long, small diameter sampling lines to the location of the analysis. The result of the analysis is sensitive to reactions in the sampling line. Thus, errors can be introduced independent of the technique for the actual analysis. As an example, it could be mentioned that only a slight increase in the corrosion potential limits flow assisted corrosion in a secondary side environment. The ECP is very sensitive to small amounts of oxygen and very accurate monitoring techniques are needed to detect small amounts of oxygen and related ECP changes. In Ref. 5, an overview is given of monitoring applications in BWRs and in PWR primary and secondary systems focusing on the problems of monitoring low oxidant contents. The new technique presented in this paper describes an electrode arrangement for ECP measurements in PWR secondary systems to obtain an improved sensitivity with respect to the detection of oxygen.

The new technique would also be applicable for BWR measurements. There are many similarities between the PWR secondary side and the BWR environment even though the actual reactions are different. Both in a BWR on HWC and in the PWR secondary system, the corrosion potentials show a large variation between different system parts. Thus, modelling of chemical and electrochemical conditions along reactor systems is important, as actual measurements cannot be performed at every point of interest. Also, improved modelling both for BWR and PWR secondary systems has recently been performed. Such models are helpful to postulate the local chemical and electrochemical conditions.

13.2   Previous experience

To better characterise the redox conditions in PWR secondary systems, electrochemical measurements have been used in several PWRs. Measurements have been performed mainly using autoclaves connected to sampling lines. In addition to such side-stream monitoring, installations directly into the feed water piping or into steam generators have been reported.

The general purpose of the measurements was to investigate the effect of water chemistry during normal and transient conditions on the corrosion and redox potentials at different locations in the secondary side system. In particular, the response to hydrazine and to transients in the form of air ingress to the steam and the water phase has been studied. Measurements have also been performed during start-up and shut-down conditions.

Measurements in PWR secondary systems have been performed at several sites including: Palo Verde (USA), St Lucie (USA), Biblis (Germany), Ringhals (Sweden), Kewaunee (USA), Takahama (Japan), Comanche Peak (USA), Bruce (Canada) and Tsuruga (Japan). The results of these measurements show that the ECP decreases with increasing hydrazine content and decreasing oxygen. However, the decrease occurs at different concentrations in different plants. The reasons for such differences are the problems of monitoring slightly oxidising conditions in high-temperature water. These problems are well-known in BWRs and new measuring techniques to monitor ECPs in BWRs on HWC have been developed. Except for the measurements presented in Ref. 4, the PWR secondary side measurements mentioned above have all been performed using autoclaves and the actual ECP response depends most likely on the installation points for the measuring equipments. Such effects have been demonstrated clearly during measurements in the Ringhals PWRs in Sweden and those results are shown and discussed below.

13.2.1   Measurements at the Ringhals PWRs in Sweden

Measurements in the PWR secondary systems of the Ringhals reactors were initiated at the start of the 1990s, bearing the experience from measurements in BWRs in mind. Autoclave equipment was used for the first measurements and three autoclaves were installed at different points in the secondary system, see Figs. 13.1 and 13.2. The sampling lines were kept as short as possible and were not longer than 1 m.

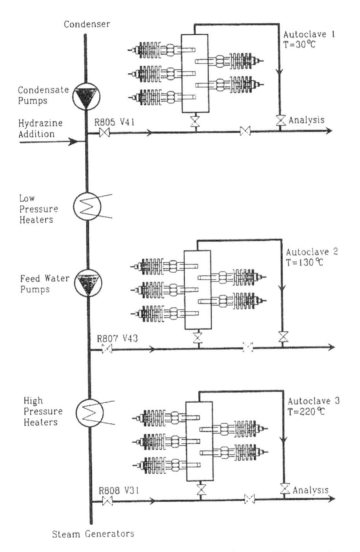

13.1 Monitoring locations in the secondary systems at Ringhals. Autoclave 1 was installed after the condensate pumps. At this location, the temperature is close to ambient. Autoclave 2 was installed after the feed water pumps and the low-pressure feed water heaters. The temperature at this location is 120–130°C. Autoclave 3 was installed after the high-pressure feed water heaters. The temperature at this location is 210–230°C

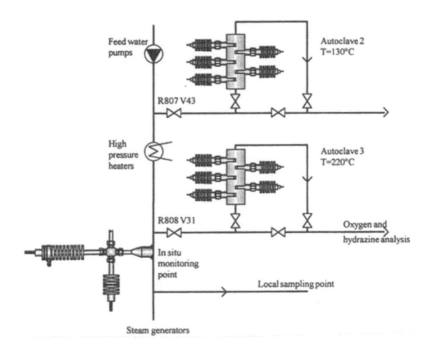

13.2 Installation point for the in-pipe electrode close to the sampling line for the third autoclave. The installation was made in a pressure monitoring gauge directly into the feed water piping

The results of the secondary side measurements at Ringhals can be summarised as follows [2,3]:

- oxidising conditions were measured at locations 1 and 2 independent of hydrazine content
- normally reducing conditions at location 3
- fast response to transients-transients not detected by other measurements were identified
- electrochemical measurements are more sensitive to redox variations than other methods used in-plant
- reliable service has been demonstrated
- electrochemical monitoring offers improved water chemistry surveillance
- interesting method for routine use.

However, an influence of the sampling lines was noted in spite of the short lines used. To avoid the sampling problem, it was decided to install measurement equipment directly into the feed water piping. A suitable measurement location (a pressure monitoring point) was located close to the autoclave sampling line, see Fig. 13.2. The seals of the electrodes were of a type previously verified in BWR measurements. The in-pipe electrode arrangement comprised a Pt electrode and a silver/silver chloride electrode.

13.2.2   Comparison between in-pipe and side-stream autoclave measurements

Examples of results are given in Figs. 13.3 and 13.4. During normal steady operation, there is only a very small difference between the two potentials but when an oxidant

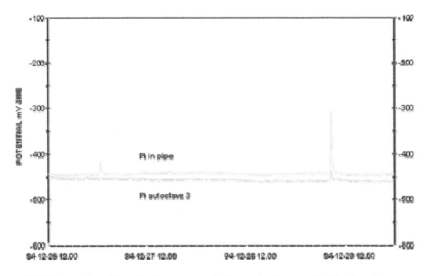

13.3 Example of transient response in the autoclave and at the in-pipe monitoring point

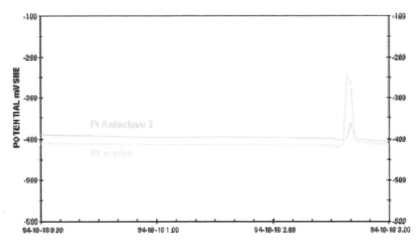

13.4 Example of transient response in the autoclave and at the in-pipe monitoring point

transient occurs, the in-pipe electrode reacts much more strongly than the electrode exposed in the autoclave. For example, in Fig. 13.3, a small oxygen transient is detected by the in-pipe measurement only (to the left in the diagram). To the right in the diagram, a larger transient occurs and the in-pipe electrode increases by 150 mV but the autoclave electrode increases by only about 25 mV. In Fig. 13.4, a similar transient occurs but here it is also clearly seen that the transient is not only smaller, but is also detected later in the autoclave compared to the in-pipe measurement.

The results of the secondary side measurements with both side-stream autoclaves and in-pipe monitoring clearly demonstrated that in-pipe monitoring is more sensitive than measurements in side-stream autoclaves. However, the necessary penetration of the primary pressure boundary of the system needed for the electrode installation is a practical drawback of the method. Safety matters with respect to potential leakages and loose parts must be treated properly which is often time consuming. (It should be mentioned that no leakages occurred during the measuring period and no parts of the electrodes were lost.)

There are two reasons for the different responses to oxygen transients as measured by the side-stream autoclave and the in-pipe electrode:

- reactions in the sampling lines
- the effect of flow-rate on ECP.

For ECP modelling in the PWR secondary system, a technique has been developed [6] which, in principle, is very similar to the earlier BWR ECP models [7]. The model can also be used to calculate the decrease in oxygen content along a sampling line. Figure 13.5 shows an example of the calculations. In Fig. 13.6, another example is shown also including the calculated ECP value along the sampling line [8].

Both Figs. 13.5 and 13.6 illustrate that the sampling line must be short to avoid sampling errors due to oxygen consumption in the sampling lines. If the distance is kept short, reasonable conditions in the sampling line can be obtained compared to the actual system.

However, to simulate the system conditions, the ECP measurements have to be performed at corresponding flow-rate. Conventional autoclaves usually have a fairly large diameter and thus the flow-rate and the mass transfer conditions are lower compared to the actual system. In Fig. 13.7, the dependence of flow-rate is illustrated.

It is concluded that the flow-rate should be high to be able to detect low levels of oxygen in the side-stream autoclave. It can be seen from Fig. 13.7 that, for an oxygen content of 0.1 ppb, no effect on the ECP is obtained at 0.1 m/s which is a typical value for side-stream autoclaves. However, a significant increase is obtained for flow-rates of 1 m/s and above, which is typical for feed water piping.

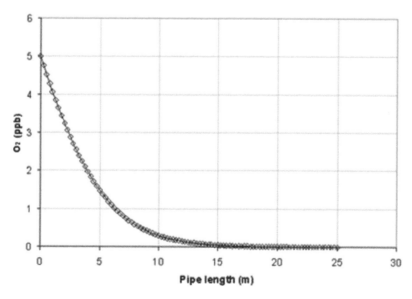

13.5 Simulated decay of oxygen in a 15 mm SS sampling line at a flow-rate of 100 kg/h, 225°C and 100 ppb of hydrazine

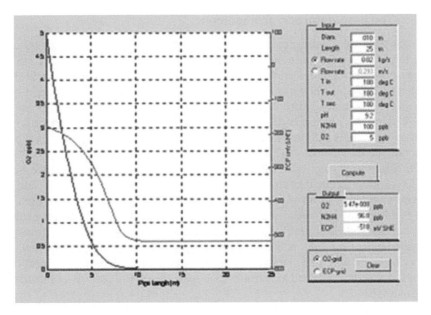

13.6 Simulation of a 25 m long, 10 mm CS sampling line at 180°C with a flow-rate of 20 g/s at 100 ppb N_2H_4 [6]

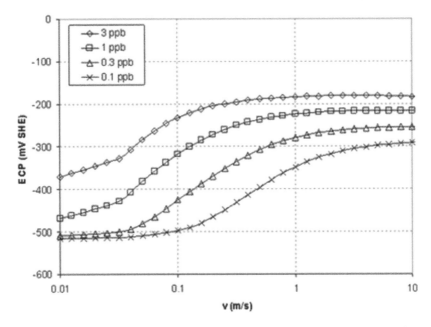

13.7 Simulated ECP of 15 mm CS sampling line at 225°C as a function of flow-rate at 100 ppb N_2H_4 and four different O_2 levels

The detection of oxygen is not only important to detect transients, it is also important to follow a deliberate addition of oxygen to the feed water system and the elimination of oxygen before the steam generators [8].

13.3   The new electrode arrangement

The new electrode design for high linear flow rate measurements in sampling lines with limited mass flow rate avoids the potential errors described above and needs no new penetration of the primary system pressure boundary. The electrode arrangement is installed as shown in Fig. 13.8. An existing sampling line is used and the T-fitting is recommended to be installed as close to the system pipe as possible. On one of the exits, the ECP device is installed and downstream of the equipment, another T-fitting is installed in the sampling line. This installation makes it possible to isolate the ECP device during power operation.

The actual ECP electrode is shown in Fig. 13.9. It comprises a reference electrode and a working electrode. The reference electrode is placed in a tube, and in the annular channel, high flow-rates are obtained. The reference electrode used is of a proven design and has been used for more than 20 years in nuclear power plants. The working electrode design provides a realistic linear flow-rate and mass transfer rate at limited mass flow-rate, and a linear flow-rate of up to 3 m/s at a mass flow-rate of 100 kg/h can be obtained. The selected flow-rate should reproduce the mass transfer rate of the system studied to ensure that realistic ECP data are obtained at low oxygen levels.

The advantages of the new technique are summarised as follows:

- The sensor can be installed without penetration of the pressure boundaries of the system.

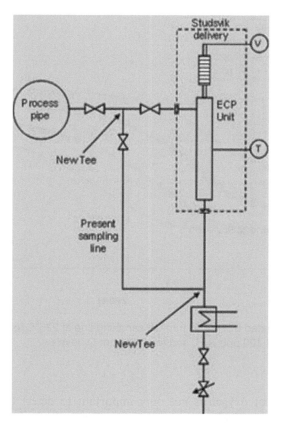

13.8 Installation of the ECP device

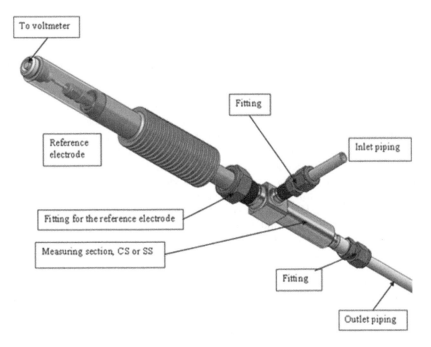

To voltmeter

Fitting

Reference electrode

Inlet piping

Fitting for the reference electrode

Measuring section, CS or SS

Fitting

Outlet piping

13.9 Electrode design

- Existing sampling lines can be used.
- The sensor comprises well-known parts that have been used for in-plant measurements in several reactors over many years.
- Local water chemistry and corrosion conditions can be accurately monitored and measurements can also be applied to flow-accelerated corrosion related studies.

The first in-plant experience with the new electrode arrangement was recently published and the results are given in Ref. 9.

13.4   Conclusions

ECP monitoring is beneficial for feed water surveillance. A new ECP sensor has been developed for the PWR secondary system offering more accurate measurements with higher sensitivity to residual amounts of oxygen. The sensor allows measurements corresponding to high linear flow-rates of up to 3 m/s at a mass flow-rate of 100 kg/h. The first in-plant application has been published. In addition, tools for combined experimental and modelling efforts are available to increase the knowledge of secondary side chemistry.

References

1. A. Molander, 'Corrosion potential monitoring in nuclear power environments'. Keynote lecture at EUROCORR'04, Nice, France. To be published by European Federation of Corrosion.

2. A. Molander et al., 'Electrochemical measurements in secondary system of Ringhals 3 PWR', in *Proc. Sixth Inter. Conf. on Water Chemistry of Nuclear Reactor Systems*, Bournemouth. BNES, 1992.
3. A. Molander et al., 'Studies of redox conditions in feed water of PWR secondary systems', in *EUROCORR'96*.
4. W. Beyer, B. Stellwag and N. Wieling, 'On-line monitoring of electrode potentials in the steam generator of a PWR', in *Proc. Third Int. Symp. on Environmental Degradation of Materials in Nuclear Power Systems – Water Reactors*. TMS, 1988.
5. A. Molander, 'Electrochemical measurements in nuclear power environments', in *Proc. of the 15th Int. Conf. on the Properties of Water and Steam* (Berlin, 7–11 September 2008).
6. M. Ullberg, P.-O. Andersson, K. Fruzetti and H. Takiguchi, 'Oxygen scavenging by hydrazine in PWR secondary feedwater – An electrochemical approach', in *Proc. Int. Conf. on Water Chemistry of Nuclear Reactor Systems* (Berlin, 15–18 September 2008).
7. A. Molander and M. Ullberg, 'The corrosion potential of stainless steel in BWR environment – Comparison of data and modeling results', in *Proc. Symp. on Water Chemistry and Corrosion in Nuclear Power Plants in Asia 2003*. Atomic Energy Society of Japan, 2003.
8. H. Takiguchi, E. Kadoi and M. Ullberg, 'Study on application of oxygenated water chemistry for suppression of flow assisted corrosion in secondary system of PWRs', in *Proc. of the 14th Int. Conf. on the Properties of Water and Steam* (2004).
9. H. Takiguchi, 'Oxygenated water chemistry for PWR secondary system – New approach to FAC, in *Proc. Inter. Conf. on Water Chemistry of Nuclear Reactor Systems* (Berlin, 15–18 September 2008).

14

Precipitation and transformation of iron species in the presence of oxygen and hydrazine in a simulated stainless steel feed water system

Jerzy A. Sawicki

AECL, Chalk River Laboratories, Chalk River, Ontario, K0J 1J0 Canada
Interatomics, Victoria, British Columbia, V8Z 7X8 Canada
jasawicki@shaw.ca

Anders Molander

Studsvik Nuclear AB, S-61182 Nyköping, Sweden

Agnès Stutzmann

Électricité de France, CEIDRE – Département Etudes, F-93206 Saint Denis, France

14.1   Introduction

In the secondary water circuit of pressurised water reactors (PWRs), colloidal particles of oxidised, mostly iron-bearing corrosion products are transported with feed water to the steam generators (SG). Iron-bearing corrosion products originate from various sites in the secondary cooling system, and especially from the steam condensers. The sludge accumulated in SGs may cause a variety of problems, such as a decrease in heat-transfer capability through deposition on tubes, limitation of flow in some restricted areas (such as tube-to-tube-support plate crevices), and acceleration of corrosion in crevices, either in deep sludge piles or at blocked tube supports. The influx of oxidised corrosion products may have a particularly adverse effect on the redox environment of SG tubing, thereby increasing the probability of localised corrosion (e.g. intergranular attack and intergranular stress corrosion cracking). One way to reduce susceptibility to intergranular attack may be to minimise the ingress of oxidised species such as ferric (Fe^{3+}) oxides and oxy-hydroxides, especially haematite (α-Fe_2O_3), goethite (α-FeOOH) and lepidocrocite (γ-FeOOH). Under reducing conditions, these species can be partly reduced to magnetite (Fe_3O_4), which is increasingly considered in the industry as a convenient indicator of acceptable redox conditions in feed water [1,2].

To suppress the formation and transport of oxidised corrosion products to the SGs, and to minimise corrosion of the SG tubes, an all-volatile treatment (AVT) of feed water is commonly used in PWR plants. It consists of ammonia or ethanolamine as a pH control agent and hydrazine, and sometimes carbohydrazide, as oxygen scavengers. Hydrazine is a particularly strong oxygen scavenging agent at pH 9 to 10.5 in an operating temperature range of 100 to 275°C because of the oxidation reaction $N_2H_4 + O_2 \rightarrow N_2 + 2H_2O$ [3–6], but below 100°C, the rate of this reaction is low. The decomposition reaction $3N_2H_4 \rightarrow N_2 + 4NH_3$ becomes fast above 275°C, which would lower the effectiveness of oxygen removal at SG temperatures. Hydrazine also inhibits corrosion at ambient temperatures and is often used for the protection of carbon steel components under shutdown conditions. Evans [4] proposed that the inhibitive action of hydrazine at ambient temperatures is caused by hydrazine replacing the aggressive anodic reactions $Fe \rightarrow Fe^{2+} + 2e^-$ by the harmless anodic reaction $N_2H_4 + 4OH^- \rightarrow N_2 + 4H_2O + 4e^-$. According to Lowson [5], hydrazine at concentrations higher than 0.01 mol/l provides complete corrosion protection by the reaction $N_2H_4 \rightarrow N_2 + 4H^+ + 4e^-$. Hydrazine may also be consumed in direct interaction with reducible iron oxides and oxy-hydroxides, especially due to heterogeneous processes at oxidised system surfaces. These processes are catalytic in nature and their rates are usually not known for complex feed water systems. The overall impact of heterogeneous reactions seems to be underestimated in the models and discussions. We still lack important data and fundamental knowledge on oxygen reduction by hydrazine, the influence of AVT chemistry on corrosion product transport and the influence of crud transport and transients on SG tubing corrosion. In particular, the ability of different hydrazine concentrations to reduce oxidised iron species dissolved in feed water to magnetite is not satisfactorily understood and quantified.

To ensure a low degree of oxidation of corrosion products transported to SGs, the PWRs often tend to operate at excessively high concentrations of hydrazine, which is not really based on good understanding and may even be harmful, particularly from the point of view of enhanced flow assisted corrosion (FAC), in highly deoxygenated, carbon steel systems. For example, some Japanese PWRs have operated in the past with as much as 600 µg/l feed water hydrazine, based on the observation that haematite was not found in SG blowdown water in plants operating at such high hydrazine concentrations. The Electric Power Research Institute (EPRI) Secondary Water Chemistry Guidelines recommend operating with greater than 100 µg/l feed water hydrazine, and as much as 90% of magnetite in the SG blowdown as a target value [7]. Électricité de France specifications for copper-free PWR plants [8] and AECL specifications for CANDU PHWR plants [1] recommend feed water hydrazine >50 µg/l, and >5 µg/l in case of copper-bearing feed trains. German plants operate between 20 and 70 µg/l feed water hydrazine concentrations and at elevated pH. Finding the underlying data to optimise concentration of hydrazine, which would be sufficient for ensuring required reducing conditions in feed water and SGs while preventing excessive FAC [9], is important both for economic and environmental reasons.

In this work, the distribution of iron redox states along a simulated feed water train was studied at different feed water chemistries, which have been set up essentially by changing concentrations of oxygen and hydrazine and keeping pH 9.8. The goal was to study the relationship between the electrochemical potentials measured in feed water and the composition of the sampled iron oxide and oxy-hydroxide particles at various temperatures and chemical conditions. In-situ electrochemical measurements, as well as ex-situ X-ray fluorescence and Mössbauer spectroscopy analyses of filtrates were extensively used. Preliminary results of this work were presented briefly elsewhere [10].

14.2   Experimental techniques

The experiments were performed in a special loop, simulating conditions from condensate to final feed water. Three pipe-shaped autoclaves of stainless steel were connected in series, as shown in Fig. 14.1. The volume of each autoclave was approximately 0.3 l and the water flow rate was 50 l/h. The transport times were estimated to be approximately 3 min between the autoclaves, whereas the total transport time was about 7 min. The velocity of water in each autoclave was estimated to be 0.5 cm/s. Heaters were used between the autoclaves to obtain representative temperatures. The water in the first autoclave (autoclave 1) was at ambient temperature, to simulate PWR condensate conditions. In the second autoclave (autoclave 2), the water temperature was 185°C, as in the low-pressure preheaters, and in the third autoclave (autoclave 3) the water temperature was at 235°C, as in the high-pressure preheaters. The temperatures and pressures were chosen on the basis of conditions in French 1300 MW PWRs, such as in Golfech 2 [11].

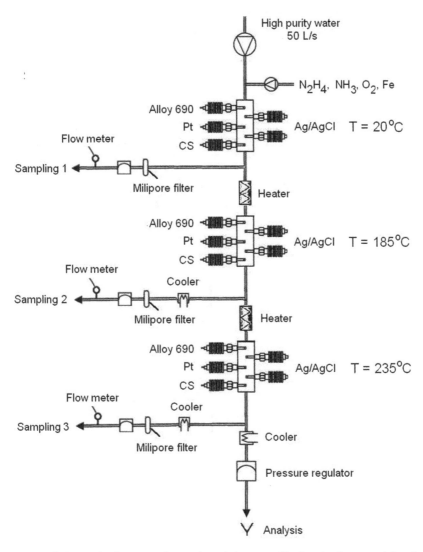

14.1 Schematic diagram of experimental set-up. Each autoclave contains 3 working electrodes and 2 reference electrodes

Each autoclave contained 3 air-cooled electrodes with cylindrical platinum, Inconel 690 and carbon steel tips, as well as 2 air-cooled silver–silver chloride reference electrodes of external type without addition of KCl. The temperature and electrochemical potential (ECP) of the autoclaves (stainless steel 316) were also recorded. The values of ECPs are given in reference to standard hydrogen electrodes (SHE) at the corresponding temperatures. The methodology of ECP measurements performed in this work was the same as that used in Ringhals 3 PWR and described by Molander *et al.* [12].

The tests lasted 17 days. The values of various chemistry control parameters recorded at the inlet and outlet of the test loop throughout the duration of the tests are listed in Table 14.1. Figure 14.2 gives a graphic representation of the variations in pH, hydrazine and oxygen at the inlet of the loop. The electrochemical records are summarised in Fig. 14.3.

Additions of iron (10 µg/kg, except tests 15 to 17, where the addition was 1 µg/kg), ammonia (to pH 9.8, except tests 6 to 8, where ammonia was added to pH 10.2), hydrazine (from 0 to 1 mg/kg) and oxygen by air-saturated water (5 µg/kg, except tests 1 to 3, where the addition was 10 µg/kg) were made to autoclave 1. Additions of iron were made as iron(II) ammonium sulphate, $(NH_4)_2Fe(SO_4)_2$, dissolved in de-aerated water.

The outlet hydrazine concentration was measured with a spectrophotometer using Merck Spectroquant 14797 reagent. The outlet oxygen concentration was measured using an Orbisphere oxygen analysing instrument. The outlet oxygen concentration was below 1 µg/kg at all dosages of hydrazine. When hydrazine was not added, the outlet oxygen concentration was 2 µg/kg (the oxygen dosage was 5 µg/kg). The conductivity of outlet water was measured at ambient temperature with a Metrohm instrument and a flow-through cell with a cell constant of 0.05 cm⁻¹. The outlet

Table 14.1   Variation of chemistry parameters during the experiment

Test	Time	NH$_3$ (ppm) intended	pH inlet	N$_2$H$_4$ (ppb) inlet	N$_2$H$_4$ (ppb) outlet	O$_2$ (ppb) inlet	O$_2$ (ppb) outlet	N$_2$H$_4$/O$_2$ inlet	Fe (ppb) inlet	pH calc outlet
1	3.26	0	9.8	50	<20	10	<1	5	10	7.6
2	3.27 am	3	9.8	50		10	1	5	10	9.7
3	3.27 pm	3	9.8	50		10	1	5	10	9.7
4	3.28	3	9.8	200		5	1	40	10	9.7
5	3.29	3	9.8	200	130	5	1	40	10	9.7
6	3.30	10	10.2	1000	900–1000	5	1	200	10	10
7	3.31	10	10.2	1000	900–1000	5	1	200	10	10
8	4.01 am	10	10.2	1000	900–1000	5	1	200	10	
9	4.01 pm	3	9.8	500		5	1	100	10	9.7
10	4.02	3	9.8	500	400	5	1	100	10	9.7
11	4.03	3	9.8	200		5	1	40	10	9.7
12	4.04	3	9.8	200	130	5	1	40	10	9.7
13	4.06	3	9.8	50		5	1	10	10	9.7
14	4.08	3	9.8	0		5	2	0	10	9.7
15	4.09	3	9.8	50		5	1	10	1	9.7
16	4.10	3	9.8	50	<20	5	1	10	1	9.7
17	4.11	3	9.8	100		5	1	20	1	

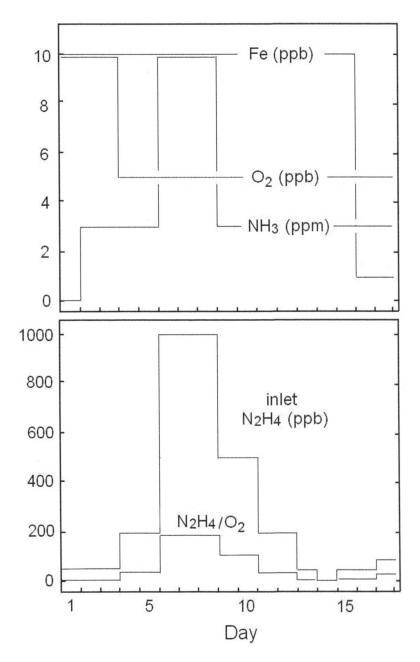

14.2 Variation of chemistry parameters during the experiment. The data include inlet pH, inlet ammonia concentration, inlet and outlet hydrazine and oxygen concentrations, as well as inlet iron concentration

conductivity was 17 to 19 μS/cm at additions of ammonia to pH 9.8 and 32 to 36 μS/cm at additions of ammonia to pH 10.2.

Solid forms of iron in the water were sampled downstream from each autoclave on 47 mm-diameter Millipore Isopore (polycarbonate) filters with a pore size of 0.2 μm.

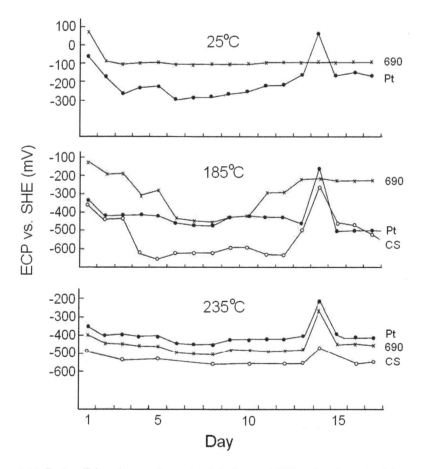

14.3 Redox (Pt) and corrosion potentials (Inconel 690 and carbon steel) in autoclaves 1, 2 and 3 for different additions of hydrazine and ammonia during the iron distribution tests. The water flow rate was 50 l/h

The integrated flow was measured with a Durant integrating flow counter. Through each filter, up to 5 l/h was drawn during sampling. The total sampled volume was 50 to 100 l. The filters were usually changed at 24 h time intervals. Exceptions were filters 2 and 8, which were exposed for 14 h, as well as filters 3 and 9, which were exposed for 10 h.

The content of Fe on the filters was determined by X-ray fluorescence analysis (XRF), using a set of reference samples and a method described elsewhere [13,14]. The XRF analyser was equipped with a 30 mCi ^{108}Cd source and a high-resolution Si-detector of X-rays. The obtained concentrations of Fe species, in μg Fe/kg H_2O, are listed in Tables 14.2 to 14.4, together with other experimental data. They are also presented in Fig. 14.4.

The chemical form and oxidation states of iron in the samples were determined by Mössbauer transmission spectroscopy using the 14.4-keV γ-rays in ^{57}Fe, and the procedure described elsewhere [15]. The application of this technique to investigations on corrosion-product transport (CPT) and redox conditions in the secondary system of PWR and CANDU reactors has been reported in more detail previously [1,16,17].

Table 14.2 Iron concentrations, percentages of various iron species, and electrode potentials in autoclave at 25°C

Test	H_2O (l)	Fe (μg/l)	Fe_3O_4 (%Fe)	α-Fe_2O_3 (%Fe)	α-FeOOH (%Fe)	γ-FeOOH (%Fe)	I690 ECP (mV)	CS ECP (mV)	Redox potential Multeq
1	51.0	6.2	–	3	–	97	74	–380	–370
2	43.8	1.8	–	2	–	*98	–85	–513	–500
3	15.3	8.9	–	2	–	*98	–104	–505	–500
4	30.1	9.8	–	3	–	*97	–100	–512	–520
5	49.8	6.9	–	7	–	*93	–99	–514	–520
6	94.2	8.9	–	5	–	*95	–103	–545	–560
7	50.9	9.3	–	5	–	*95	–102	–547	–560
8	52.2	8.8	–	4	–	*96	–99	–549	–560
9	27.8	8.7	–	5	–	*94	–102	–530	–530
10	77.1	9.3	–	5	–	*95	–97	–522	–530
11	96.3	9.4	–	4	–	*96	–83	–532	–520
12	103.1	8.8	–	6	–	*94	–85	–532	–520
13	155.2	8.6	–	5	–	*95	–83	–536	–500
14	211.4	8.2	–	6	–	*94	–95	–543	–590
15	88.3	1.4	–	38	–	*62	–92	–540	–500
16	100	1.4	–	30	12	*58	–83	–534	–500
17	98.0	1.1	–	27	7	*66	–84	–533	–510

* Very small particles of α-Fe_2O_3 or ferrihydrite.

Table 14.3 Iron concentrations, percentages of various iron species, and electrode potentials in autoclave at 185°C

Test	H_2O (l)	Fe (μg/l)	Fe_3O_4 (%Fe)	α-Fe_2O_3 (%Fe)	α-FeOOH (%Fe)	γ-FeOOH (%Fe)	Pt ECP (mV)	I690 ECP (mV)	CS ECP (mV)	Redox potential Multeq
1	59.0	4.6	6	19	57	18	–337	–124	–349	–400
2	45.6	4.8	12	15	68	5	–426	–192	–435	–490
3	15.5	5.6	7	18	70	5	–417	–188	–433	–490
4	33.8	4.1	14	7	76	1	–413	–306	–620	–520
5	49.3	3.7	10	15	69	6	–420	–276	–655	–520
6	96.2	4.2	53	0	35	1	–457	–443	–624	–580
7	49.3	6.0	60	0	39	1	–463	–455	–625	–580
8	48.9	5.7	60	0	38	2	–464	–457	–624	–580
9	27.6	5.0	54	0	43	3	–426	–426	–594	–540
10	76.8	6.2	16	43	39	2	–426	–424	–591	–540
11	88.9	6.2	10	30	58	2	–429	–291	–624	–520
12	102.6	6.2	11	25	61	3	–423	–282	–639	–520
13	146.6	6.2	6	24	67	3	–458	–222	–497	–490
14	185.7	6.0	6	20	70	4	–158	–222	–260	390
15	78.0	5.1	7	40	51	2	–499	–226	–446	–490
16	98.5	0.1	5	56	30	9	–497	–232	–473	–490
17	99.1	0.6	7	39	47	7	–499	–233	–512	–510

Note: in samples 6 to 13, magnetite seems to be either non-stoichiometric or may contain some maghemite; its fraction correlates with hydrazine concentration.

Table 14.4 Iron concentrations, percentages of various iron species, and electrode potentials in autoclave at 235°C

Test	H_2O (l)	Fe (µg/l)	Fe_3O_4 (%Fe)	α-Fe_2O_3 (%Fe)	α-FeOOH (%Fe)	γ-FeOOH (%Fe)	Pt ECP (mV)	I690 ECP (mV)	CS ECP (mV)	Redox potential Multeq
1	51.8	3	10	76	10	4	−346	−394	−486	−420
2	46.1	3.3	5	61	27	7	−398	−447	−539	−500
3	14.8	6.7	47	35	14	4	−395	−447	−538	−500
4	22.8	3.9	34	50	11	5	−409	−460	−535	−530
5	47.1	1.3	46	16	33	5	−408	−463	−534	−530
6	94.4	3.3	63	12	21	4	−448	−496	−551	−590
7	47.0	3.4	71	12	14	3	−451	−501	−556	−590
8	49.5	3.6	80	8	8	4	−452	−504	−559	−590
9	27.4	4.3	71	–	23	6	−420	−479	−545	−550
10	80.1	3.4	65	12	18	5	−427	−484	−553	−550
11	82.0	3.8	64	21	7	8	−416	−480	−551	−530
12	104.9	3.9	53	20	23	4	−417	−481	−554	−530
13	143.7	4.0	46	19	30	5	−401	−476	−551	−500
14	187.9	10.9	22	33	45	2	−213	−255	−468	300
15	53.8	1.0	16	61	22	1	−398	−442	–	−500
16	107.8	1.0	14	42	35	9	−403	−455	−561	−500
17	96.7	1.0	18	18	52	12	−416	−467	−560	−520

Note: magnetite seems to be either non-stoichiometric or may contain some maghemite; its fraction correlates with hydrazine concentration and low haematite fraction.

The analysis was performed using a 30 mCi single-line ^{57}CoRh source. The spectra of all samples were obtained at room temperature, under identical experimental conditions. Several samples were also examined at 77.3 K and 4.2 K. The obtained spectra were analysed using a least-squares fitting procedure to determine the percentage of iron in different chemical forms, i.e. iron oxides and oxy-hydroxides, as listed in Tables 14.2 to 14.4. The results of Mössbauer analyses are also presented in the form of bar charts in Fig. 14.5.

14.3   Results and discussion

14.3.1   Hydrazine consumption measurements

As the data in Table 14.1 indicate, at 10 µg/kg dissolved oxygen and 10 µg/kg added Fe, the concentration of outlet hydrazine was generally lowered by ~50 µg/kg, compared to the inlet hydrazine concentration. This value suggests that a considerable part of hydrazine had to be consumed in reactions other than the simple oxidation reaction $N_2H_4 \rightarrow N_2 + 2H_2O$. Decomposition to ammonia $2N_2H_4 \rightarrow 2NH_3 + H_2$ and heterogeneous reactions on materials surfaces are thought to be responsible for the remaining part of the observed hydrazine consumption in the loop. When hydrazine was not added (test 14), some oxygen was also consumed (from 5 to 2 µg/kg).

14.3.2   Redox and corrosion potentials

Figure 14.3 compares the variation in the potentials of platinum, Inconel 690, and carbon steel in autoclaves 1, 2 and 3, which were recorded during the experiment. The

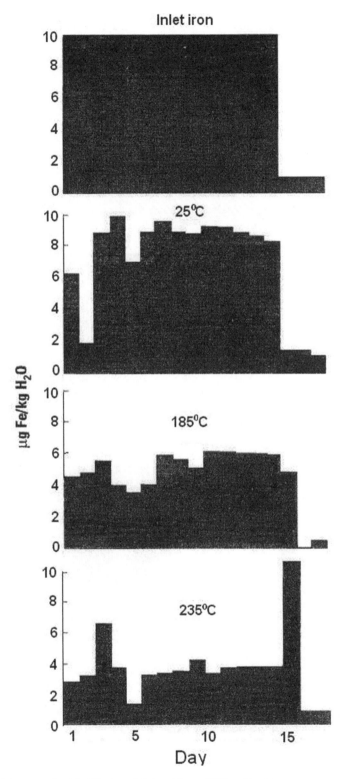

14.4 Iron content in filtrates obtained from XRF analyses

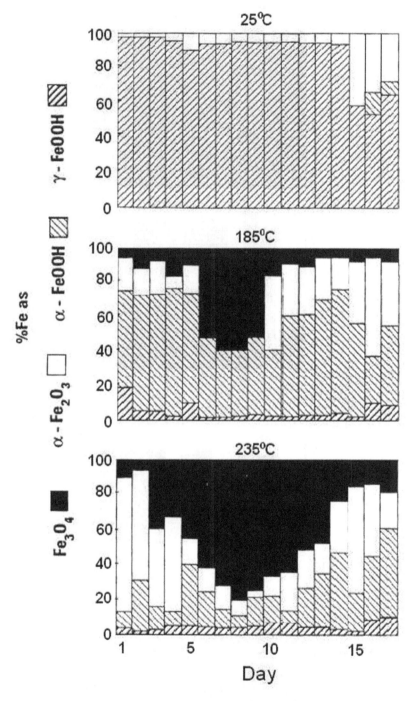

14.5 The distribution of iron oxide/hydroxide species in filtrates collected after autoclaves 1, 2 and 3 during the experiment

values of potentials averaged over each filtration period are given in Tables 14.2, 14.3 and 14.4, together with other experimental data obtained in this experiment.

The inspection of the data in Fig. 14.3 allows for the following observations to be made:

Significantly lower potentials were observed when ammonia was added; cf. test 1 (without ammonia) vs. tests 2 to 17 (ammonia added). This observation is reasonable because the potentials for the electrochemical equilibrium between water and hydrazine, between water and oxygen, and between water and ammonia are lowered at higher pH values.

Significantly higher potentials, of up to 200 to 300 mV, were observed for Pt and Inconel electrodes when hydrazine was not added; see test 14. The carbon steel electrode did not react to the lack of hydrazine at 25°C but responded slowly at 185 and 235°C. Slow response was also noted at 235°C for Pt and Inconel 690 electrodes.

Reactions to changes in hydrazine were especially strong at 185°C, for carbon steel and Inconel 690. This finding reflects corrosion conditions at mixed potentials and far from equilibrium. It is worth noting that the carbon steel electrode reached the same lowest potentials at 200 μg/kg and at all higher hydrazine additions, whereas for Inconel 690, dosages of 500 to 1000 μg/kg were required.

Further, at 235°C, all three electrodes reacted quite similarly to changes in pH and hydrazine, and their potentials differed by approximately 50 mV.

Spikes in the potential occurred during or shortly after changes in hydrazine concentration, especially during the first increases to 200 and 1000 μg/kg.

14.3.3 Transport of particulate iron

Figure 14.4 shows the variation in the iron concentration, as obtained from XRF analyses of filters. The following observations are to be noted:

Usually up to 80 to 90% of added Fe was recorded in a particulate form on filters sampled at 25°C. A very small fraction of 20% was found at the start of ammonia addition (test 2); this value can be explained possibly by a higher fraction of soluble Fe at this time.

In samples collected at 185°C, approximately 40 to 60% of Fe was recorded as particulates when added Fe was kept at the level of 10 μg/kg. However, when the dissolved Fe concentration was lowered to 1 μg/kg, the amount of particulate Fe remained quite high at first in test 15 and dropped to very low values in tests 16 and 17. This finding suggests that, on decline of Fe supply, some of the particles deposited in the system were entrained.

In samples taken at 235°C, the particulate Fe decreased generally to about 30 to 40% of the initial level, with the exception of test 15, where it increased to above the 10 μg/kg level, similar to the trend observed at 185°C. This effect may suggest again that a decreased supply of dissolved Fe did force entrainment of particulates that were previously deposited along the feed train. The amount of Fe trapped in the loop during the experiment appears to be notably small. From the sampling data obtained from autoclave 3, it can be estimated that only about 50 to 60 mg of Fe did not arrive at the outlet of the loop in the whole test. This amount of deposited crud in the system is large enough to explain the crud burst observed in test 15. Also, it is feasible that a fraction of Fe has penetrated through the filters in a dissolved or colloidal form (particle size <0.2 μm).

14.3.4   Mössbauer spectroscopy data

Figure 14.5 shows how the distributions of different iron redox states in filtrates taken from autoclaves 1, 2 and 3 evolved during the experiment. The following observations and comments may be made regarding the interpretation of the Mössbauer spectra:

The XRF and Mössbauer data for deposits collected after the first autoclave indicate that practically all Fe introduced at the inlet of the loop in the form of dissolved Fe(II) species was rapidly precipitated in the form of filterable particulates. Although these colloidal particles already have some features of individual phases, their identity is, at this level of precipitation, more amorphous than crystalline, or quasicrystalline.

The Mössbauer spectra of these freshly precipitated deposits exhibited mostly quadrupole-split doublets with an isomer shift (δ) of 0.26 mm/s and quadrupole splitting (ΔE_Q) in the range from 0.58 to 0.73 mm/s. Although these parameters were ascribed to lepidocrocite (γ-FeOOH), the presence of amorphous ferrihydrite ($Fe_5HO_8\cdot 4H_2O$) particles, akageneite (β-FeOOH) particles, and very fine superparamagnetic particles of haematite (α-Fe_2O_3) in these deposits cannot be excluded (especially in samples with $\Delta E_Q \sim 0.7$ mm/s). Filters 15 to 17, which were obtained after lowering the hydrazine concentration to zero, showed in addition a sizable fraction (30 to 40% Fe) in the form of larger particles of haematite and some fraction (–10% Fe) of yet poorly defined goethite (α-FeOOH).

In the filtrates obtained after autoclaves at 185 and 235°C, we have observed a remarkable degree of interconversion of the iron oxides and oxy-hydroxides. Such transformations in aqueous suspensions usually take place through a dissolution–recrystallisation mechanism at elevated temperatures and yield aggregates of oriented microcrystals from each reacting crystal [18].

In the filtrates obtained from the autoclaves at 185°C, the dominant species appeared to be small particles of goethite. The spectral attributes of these particles indicated large line broadening and a reduced hyperfine magnetic field, typical for the interaction of small particles of goethite. We also observed a considerable fraction of Fe in the form of magnetite (Fe_3O_4) particles; the magnetite fraction increased up to 60% in tests of hydrazine dosing to 500 and 1000 μg/kg. Magnetite observed here was rather poorly defined because it indicated an anomalously high occupancy of the tetrahedral sub-lattice; as seen in Table 14.3, the tetrahedral-to-octahedral occupancy ratio (A/B) is much higher than that usually observed in crystalline forms of magnetite.

Filtrates obtained from the autoclave at 235°C indicated further growth of the magnetite fraction (see Table 14.4) clearly coinciding with the higher concentrations of hydrazine.

14.3.5   Iron redox states at different hydrazine concentrations

As the data in Table 14.2 and Fig. 14.6 indicate, all of the iron, added as dissolved ferrous iron, was rapidly oxidised to ferric iron species, at conditions that corresponded to those of the condensate. Some poorly defined magnetite was formed

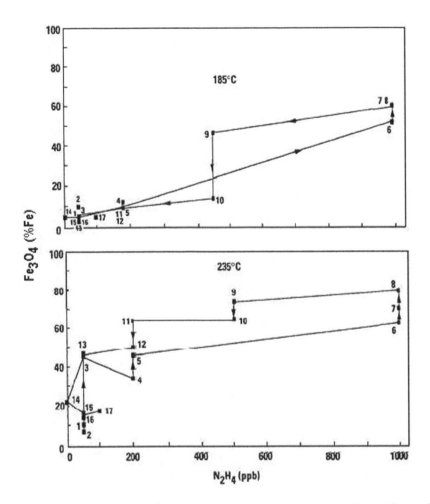

14.6 Fraction of magnetite in feed water conditions corresponding to those of low-pressure preheaters (top) and the high-pressure preheaters (bottom) for different hydrazine dosages

under conditions corresponding to those of the low-pressure preheaters. Additional and better-defined magnetite was formed at conditions corresponding to those of the high-pressure preheaters.

It is also worth noting that for the test without ammonia but with 50 µg/kg hydrazine, the fraction of magnetite was lower than for the test condition without hydrazine but with ammonia at pH 9.8 (compare tests 1 and 14 in Fig. 14.6). This finding demonstrates that not only the hydrazine level but also the pH influences the reduction/oxidation state of the iron species.

Figure 14.6 shows the fraction of magnetite for the different feed water conditions (at conditions corresponding to those of the low-pressure and high-pressure preheaters, respectively) at different hydrazine concentrations.

The increasing fraction of magnetite for increasing hydrazine dosage, as observed in the final feed water, clearly demonstrates the beneficial effect of hydrazine in forming magnetite from trivalent iron species. It is notable that the fractions of magnetite obtained here are in the same range as in plant data [1,2,17–20]. Although the laboratory and plant data of Nakamura *et al.* [19] show a large scatter in the fraction of magnetite at lower hydrazine concentrations, a tendency for the fraction of magnetite to increase at increasing hydrazine concentrations was noted. The PWR plant data of Sawochka *et al.* [20] and CANDU data reported by Sawicki and colleagues [1,17] showed a decreasing trend of magnetite for higher hydrazine concentrations. The reason for this decreasing trend is not clear. However, the results are within the scatter described by Nakamura *et al.* [19].

Thus both plant data and present laboratory data indicate that, even at high hydrazine dosages, full conversion to magnetite is not reached in the feed water of the high-pressure preheaters. However, data on fractions of haematite in feed water presented by Hino *et al.* [21] show that as little as about 5% of haematite could be observed in some Japanese PWR plants at hydrazine concentrations above approximately 200 µg/kg. This finding is in contrast to US PWR plants and CANDU reactors.

A comparison of magnetite fractions in SG blow down water from different PWR and CANDU power stations has been given by Sawicki *et al.* [1]. The trend of increasing fraction of magnetite at increasing hydrazine to dissolved oxygen ratio was observed in the blow down samples. However, blow down water has been exposed to higher temperatures for a longer time than the feed water, which favours magnetite formation. Thus, this water cannot be used to diagnose the corrosion conditions of the water at the inlet to the SG.

14.3.6   Iron redox states at different potentials

Figure 14.7 shows the fraction of magnetite in feed water corresponding to conditions of the high-pressure preheaters for different potentials of platinum, Inconel 690 and carbon steel electrodes. As seen, for all tests where ammonia was added except test 1, a lower potential of platinum and Inconel 690 is correlated with a higher fraction of magnetite under the feed water conditions corresponding to those of the high-pressure preheaters. Hydrazine has a reducing effect, and thus higher hydrazine concentrations lead to lower potentials, and increase the reduction of ferric iron species in the feed water to magnetite. However, as seen in Fig. 14.7, the corrosion potential for carbon steel was in this case generally low, and the correlation between the fraction of magnetite and the potential is not as clear as it is for platinum and Inconel 690. The results also show that the fraction of magnetite increases with decreasing potentials of platinum and Inconel 690 at 235°C, where ammonia is present.

The proportion of observed magnetite may be correlated with the Pt electrode potential through the Pourbaix diagrams for the $Fe-H_2O$ system at ambient and feed water temperatures [22]. At 25°C, the potentials, both measured (see Table 14.2 and Fig. 14.3) and calculated with the Multeq code (see discussion below), are outside the region of stability of Fe_3O_4. At this temperature, pH and potentials, we are in the region of the diagram where Fe_2O_3 is stable, but due to close proximity of the field of dissolved $Fe(OH)_2^-$ species, metastable γ-FeOOH are formed preferentially. On the other hand, at 185°C and at higher temperatures, a potential below -500 mV is more favourable for Fe_3O_4 (pH at 185°C ~ 7) formation, and less favourable for the existence of ferric iron precipitates.

14.3.7   Comparison of measured and calculated redox potentials

The potential of electrodes in flowing water systems is a kinetically-controlled electrochemical parameter. Due to lack of all of the kinetic parameters required for

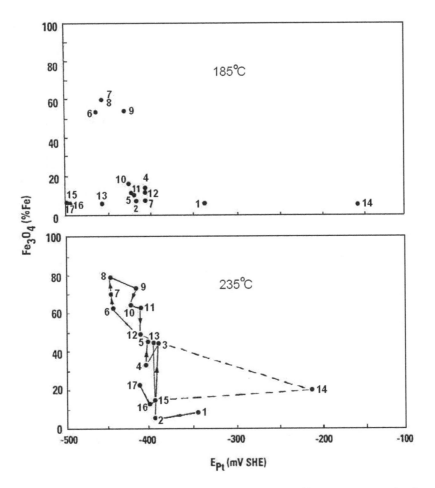

14.7 Fraction of magnetite in feed water conditions of low-pressure preheaters (top) and high-pressure preheaters (bottom) as a function of potential of platinum electrodes

the description of complex interactions occurring in the carbon steel-AVT secondary water system, this potential cannot be modelled at present with any degree of accuracy using, for instance, a model of mixed potentials that had been applied by MacDonald and Urquidi-Macdonald [23] in the description of the electrochemical conditions under normal or hydrogen water chemistry in boiling water reactors (BWRs).

As the first step in modelling redox potentials in PWR feed water, we have attempted to use the Multeq code [24], which calculates redox potentials using an extensive database, but only at equilibrium conditions. We wanted to find out how the potentials measured in this work would compare with equilibrium redox values predicted by Multeq for the sets of chemical parameters in the feed water of the loop in each of our tests. In other words, we wanted to find equilibrium boundaries to our essentially non-equilibrium situation.

A comparison of measured Pt electrode potentials and Multeq-calculated redox potentials in Fig. 14.8 indicates quite a continuous, albeit not one-to-one,

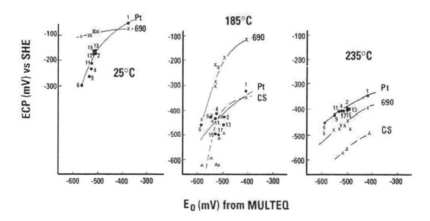

14.8 Comparison of redox potentials calculated using the Multeq code with measured ECP of platinum, Alloy 690 and carbon steel electrodes, at 25, 185 and 235°C in tests numbered from 1 to 17

correspondence between measured and calculated potentials for each electrode and temperature (even for tests 1 and 14 with no hydrazine applied). The following observations could be made:

At 25°C, the Pt electrode potential at low measured values (between –150 and –300 mV) is much more positive and changes more steeply with changes in chemistry than the calculated redox potential (between –480 and –600 mV). Its calculated values in tests 1 and 14 (–400 and 600 mV) are even more drastically different from the measured ones (–50 and 80 mV). The measured potential of Alloy 690 changes very little (within ~30 mV) with the conditions of the test and is very far from Multeq-calculated values and trends. Thus, at low temperature, Multeq does not give any prediction of redox potentials at the chemical and kinetic conditions of condensate water.

At 185°C, the measured Pt electrode potentials indicate quite a scatter of points around an average guideline, but in general, the range of measured values (–330 to –500 mV) compare quite well with the range of calculated ones (–400 to –600 mV); with the exclusion of the remote point 14 (at –200 and 420 mV, respectively). Alloy 690 measured potential changes more regularly, but more steeply (between –120 and –450 mV) than its calculated redox counterpart (–400 to –600 mV), while at point 14 it assumes a drastically unmatched value (–250 vs. 400 mV). Carbon steel potential also changes very steeply and differently compared to the calculated redox potential. Thus, the Multeq calculations indicate that at 185°C, the chemical conditions in the loop were far from equilibrium.

At 235°C, the potentials of all three electrodes correlate quite linearly with the calculated redox potentials, but vary somewhat less steeply than calculated redox potentials (with the obvious exclusion of test 14). The Pt electrode potential ranges from –350 to –450 mV and Inconel 690 potential varies between –400 and –500 mV, while calculated redox potential varies between –420 and –600 mV. Limited carbon steel data show a parallel trend, at significantly lower values. Thus, at sufficiently high temperature, such as 235°C, under the chemical conditions of our loop tests, the Multeq code can predict general trends of potential of all three types of electrodes, but that the values may be offset and that the proportionality factor between measured potentials and calculated values is about 0.5.

14.4   Summary and conclusions

We have studied the effect of hydrazine and dissolved oxygen on precipitation of dissolved iron ions and redistribution of iron redox states along a simulated feed water train of a PWR. The tests were performed in an all-stainless steel, once-through pressurised loop equipped with three autoclaves operating at temperatures of 25, 185 and 235°C, to simulate condensate, feed water and high-pressure preheaters in the French 1300 MW Golfech 2 PWR feed train. Iron, in the form of ferrous ammonium sulphate dissolved in de-aerated water, as well as ammonia, hydrazine and oxygen at chosen concentrations were co-injected at the inlet of the loop. Systematic tests were performed at dissolved oxygen concentrations of 5 or 10 µg/kg and at hydrazine concentrations of 50, 100, 200, 500 and 1000 µg/kg.

A methodology was developed of examining the redox state (population of Fe^{2+} vs. Fe^{3+} species) of transported colloidal iron-bearing particles in conjunction with ECP measurements. Platinum, Inconel 690, and carbon steel electrodes were used to monitor ECP, and integrated crud sampling combined with Mössbauer spectroscopy was used to investigate the speciation of iron particles.

A correlation between the fraction of magnetite and ECP values of platinum and Inconel 690 electrodes was observed at simulated feed water conditions in high-pressure preheaters of PWRs.

The beneficial effect of hydrazine in the formation of magnetite from ferric iron species has been demonstrated for simulated feed water conditions.

The results indicate that, even at high hydrazine dosages such as 500 and 1000 µg/kg, ferric species cannot fully convert into magnetite under the conditions corresponding to those of the high-pressure preheaters.

Comparison of measured and Multeq-calculated equilibrium redox potentials suggest that, at 235°C, the electrochemical conditions in the loop are fairly close to equilibrium.

Acknowledgements

This work has been jointly funded by Électricité de France, Studsvik Nuclear AB, AECL and CANDU Owners Group. Special thanks go to K. Pein for her contributions in the tests and to P.-O. Andersson (Vattenfall), M. Brett (Ontario Hydro) and R.L. Tapping (AECL) for discussions and valuable comments.

References

1. J. A. Sawicki, M. E. Brett and R. L. Tapping, 'Corrosion product transport, oxidation state and remedial measures', in *Third International Steam Generator and Heat Exchanger Conference*. Canadian Nuclear Society, Toronto, June 1998, 465–479.
2. A. Molander, K. Pein, P.-O. Andersson and L. Björkvist, 'Some aspects of feedwater and steam generator chemical and redox monitoring in PWR secondary system', in *Proc. Fontevraud-98*.
3. W. F. Stones, *Chem. Ind.*, 2 (1957), 120–128.
4. U. R. Evans, *The Corrosion and Oxidation of Metals*, 1st Suppl. Vol., 173. Arnold, London, 1960.

5. R. T. Lowson, *Br. Corros. J.*, 12 (1977), 175.
6. S. B. Dalgaard and M. O. Sanford, 'Review of the hydrazine/oxygen reaction kinetics', in *Materials Performance*, 32. 1982.
7. *PWR Secondary Water Chemistry Guidelines – Revision 4.* Report TR 10234. Electric Power Research Institute, October 1996.
8. P. Berge, 'PWR – what are the choices for the plant chemist?' in *Proc. 1998 Plant Chemistry meeting*, Huntington Beach, 1998 September, GC–112049, paper 1.1.
9. B. Chexal, J. Horowitz, B. Dooley, P. Millett, C. Wood, R. Jones, M. Bouchacourt, F. M. Remy, F. Nordmann, P. Saint Paul and W. Kastner, *Flow Accelerated Corrosion in Power Plants.* Report TR-106611-R1. Electric Power Research Institute, 1998.
10. K. Pein, A. Molander, J. A. Sawicki and A. Stutzmann, 'Distribution of iron redox states for different hydrazine concentrations and potentials – a laboratory study', in *Eighth Int. Symp. on Environmental Degradation of Materials in Nuclear Reactors*, Amelia Island, Florida, August 1997, Vol. 1, 113.
11. A. Stutzmann, 'Results of tests performed at Golfech 2 with a high hydrazine – ECP measurements', in *Workshop on PWR and CANDU Electrochemical Monitoring*, Toronto, June 1999, AECL-11986, COG 98-32-I, 238.
12. A. Molander, B. Rosborg, P.-O. Anderson and L. Björnkvist, 'Electrochemical measurements in secondary system of Ringhals 3 PWR', in *Sixth Int. Conf. on Water Chemistry of Nuclear Reactor Systems*, 880. British Nuclear Energy Society, London, 1992.
13. B. D. Sawicka, 'Nondestructive elemental analysis of corrosion and wear products from primary and secondary CANDU water circuits', in *Fourth Int. Conf. on CANDU Maintenance*, 234. Canadian Nuclear Society, Toronto, Canada, 1997.
14. B. D. Sawicka and J. A. Sawicki, 'Analysis of corrosion product transport using nondestructive XRF and MS techniques', in *Third Int. Steam Generator and Heat Exchanger Conference*, 641. Canadian Nuclear Society, Toronto, June 1998.
15. J. A. Sawicki and M. E. Brett, *Nucl. Instrum. Meth. Phys. Res.*, B76 (1993), 254.
16. J. A. Sawicki, M. E. Brett, J. Price and G. Kozak, 'Oxidation state of iron at elevated hydrazine levels: Bruce-B corrosion product transport study', in *EPRI PWR Plant Chemistry Meeting*, Lake Buena Vista, Florida, November 1995.
17. M. E. Brett, A. P. Quinan, J. Price and J. A. Sawicki, 'Secondary side electrochemical potential monitoring and the redox state of corrosion products in Ontario Hydro Nuclear', in *Seventh Int. Conf. Water Chemistry of Nuclear Reactor Systems*, 407. British Nuclear Energy Society, London, 1996.
18. M. A. Blesa, M. Mijalchik, M. Villegas and G. Rigotti, *React. Solids*, 2 (1986), 85.
19. T. Nakamura, T. Kusakabe, K. Takami, M. Ishibashi, T. Hattori, H. Ohta and K. Murata, 'Development of a technique to study behavior of iron oxides in PWR secondary water', in *Proc. 1991 JAIF Int. Conf. Water Chemistry in Nuclear Power Plants*, 121. Japan Atomic Industrial Forum, Fukui City, Japan, 1991.
20. S. G. Sawochka, S. S. Choi and K. Fruzetti, 'Analyses of PWR corrosion product samples for iron redox state', Presented at *EPRI PWR Plant Chemistry Meeting*, Lake Buena Vista, Florida, November 1995, EPRI TR-106179-V2.
21. Y. Hino, I. Makino, S. Yamauchi and F. Fukuda, *Thermal Nucl. Power*, 43 (1993), 185.
22. C. M. Chen and G. J. Theus, *Chemistry of Corrosion-Producing Salts in Light Water Reactors.* PRI NP-2298. Electric Power Research Institute, 1982.
23. D. D. Macdonald and M. Urquidi-Macdonald, *Corrosion*, 52 (1996), 659.
24. Multeq, Code available from Electric Power Research Institute, Palo Alto, CA, USA.

15

An electrochemical sensor array for in-situ measurements of hydrogen peroxide concentration in high-temperature water

Shunsuke Uchida and Tomonori Satoh

Nuclear Science and Energy Directorate, Japan Atomic Energy Agency, 2-4, Shirane, Shirakata, Tokai, Ibaraki, Japan

uchida.shunsuke@jaea.go.jp

Yoshiyuki Satoh

Graduate School of Engineering, Tohoku University, Sendai, Japan

Yoichi Wada

Power and Industrial Systems R&D Laboratory, Hitachi Ltd., Japan

15.1   Introduction

Reliability loss of boiling water reactors (BWRs) has often been reported due to intergranular stress corrosion cracking (IGSCC) on stainless steel of major components and piping of primary cooling systems [1]. In order to mitigate IGSCC, water chemistry, which is one of the major factors determining IGSCC [2,3], should be improved. Corrosive conditions of BWR primary cooling systems are caused by oxidants as a result of water radiolysis. The latest information has shown that corrosive conditions in BWRs are caused mainly by hydrogen peroxide (H_2O_2) [2]. In order to understand all of the phenomena associated with corrosion behaviour, application of *in-situ* measurements is one of the key technologies. However, corrosion behaviour itself is determined in a complex fashion by many water chemistry and material parameters.

One of the most common indices of corrosive conditions is the electrochemical corrosion potential (ECP) [4]. ECP is determined not only by concentrations of oxidants, e.g. O_2, H_2O_2 and other corrosive radiolytic species, but also by specimen surface conditions, e.g. electric resistance of oxide films. However, the latest studies on H_2O_2 behaviour in elevated temperature water [5–7] have concluded that a constant level of ECP in a wide H_2O_2 concentration [H_2O_2] range was due to a balance between cathodic current density of the specimen and increasing anodic current density due to oxidation of H_2O_2. In order to understand the complicated combination of parameters influencing the corrosion behaviour, each contribution to the behaviour should be evaluated by applying *in-situ* [H_2O_2] measurements and other sensors, as well as ECP sensors [8].

In this chapter, a sensor array is introduced, consisting of ECP and frequency dependent complex impedance (FDCI) sensors; their performances are discussed and then their future applications are overviewed.

15.2   Corrosive conditions

15.2.1   Control of corrosive conditions

Water radiolysis in the reactor core and its peripheral regions generates radiolytic species, e.g. O_2 and H_2O_2, which cause highly corrosive conditions in BWR primary cooling water [1]. In order to mitigate corrosive conditions in BWR primary cooling water, hydrogen is injected into the feed water (Fig. 15.1). This is easily released into steam in the core region but the concentration of OH radicals in the irradiation field of the down-comer region is sufficiently high to cause lower oxidant concentrations in the recirculation water and then to moderate corrosive conditions in the recirculation line and the lower reactor plenum (Fig. 15.2) [9]. With increasing concentrations of hydrogen injected into the feed water, $[O_2]$ measured at the sampling point decreases rapidly and, at the same time, the main steam line dose rate (MSDR) increases gradually (Fig. 15.2) [9]. The optimal hydrogen injection amount is determined to suppress $[O_2]$ without serious increase in MSDR. Concentrations of oxidants (O_2 and H_2O_2) change along the recirculation flow path so that the corrosive conditions differ between locations in BWR primary cooling systems. The only point to determine oxidant concentrations along the recirculation flow is at the end of the sampling line installed in the recirculation line, where water is cooled down and depressurised for the probe type oxygen detector.

ECP can identify corrosive conditions directly at elevated temperatures in the primary cooling water. Measured ECPs for hydrogen water chemistry (HWC) are also shown in Fig. 15.1. From the viewpoint of IGSCC mitigation, the threshold ECP has been proposed as -230 mV$_{SHE}$ [4]. However, for the optimal $[H_2]$, determined by measured $[O_2]$, the ECP was still too high to mitigate IGSCC.

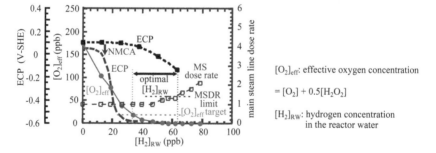

$[O_2]_{eff}$: effective oxygen concentration

$= [O_2] + 0.5[H_2O_2]$

$[H_2]_{RW}$: hydrogen concentration in the reactor water

15.1 Hydrogen injection to mitigate corrosive conditions in BWR primary coolant

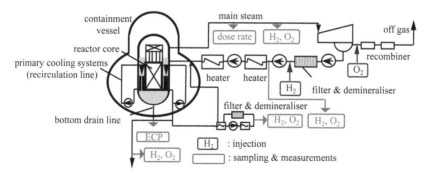

15.2 Effects of HWC on corrosive conditions

15.2.2 Theoretical evaluation

The gap between the measured ECP and measured [O$_2$] shown in Fig. 15.1 is explained by the contribution of H$_2$O$_2$ to the ECP especially at high hydrogen injection rates [10,11]. In order to determine the oxidant distribution in the primary system, theoretical models have been prepared to calculate the oxidant concentration distribution [12]. Non-linear rate equations are calculated to obtain the concentration distributions of radiolytic species throughout the primary coolant [12]. The calculated [O$_2$] and [H$_2$O$_2$] at the reactor pressure vessel (RPV) bottom are shown in Fig. 15.3 as a function of [H$_2$]$_{RW}$ in the reactor water, which shows that 100 ppb H$_2$O$_2$ coexists with 200 ppb O$_2$ under normal water chemistry conditions without hydrogen injection, while only 10 ppb H$_2$O$_2$ exists under HWC conditions with 50 ppb H$_2$ in the reactor water [13].

15.2.3 *In-situ* measurements

In order to determine [H$_2$O$_2$] and other water chemistry and material parameters as well as ECP at elevated temperatures, more than two sensors should be applied. For this purpose, a sensor array has been developed. The benefits and disadvantages of the sensor array are summarised in Table 15.1. As a first step for development of the sensor array, a flow cell type H$_2$O$_2$ detector was developed based on chemical photo-luminescence spectroscopy using a luminal chemiluminescence reaction [14]. It was applicable to measurement of sampled water at room temperature. As the second step, a combination of ECP and FDCI sensors was developed as a sensor array and their data were analysed to obtain information for determining [H$_2$O$_2$] at high temperature. As the third step, another type of sensor array has been proposed for the *in-situ* determination of the electric resistance of oxide film for laboratory

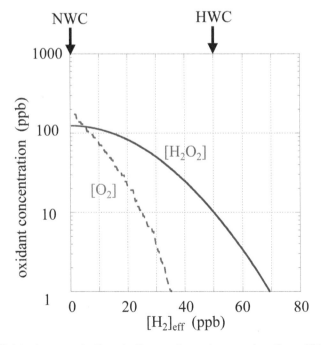

15.3 Oxidant concentrations in the reactor water as a function of [H$_2$]$_{eff}$

Table 15.1   Application of sensor array for BWR corrosive condition monitoring

	Proven type monitoring	Monitoring with sensor array
Desired	Corrosive conditions	Corrosive conditions
Measured	$[O_2]$, $[H_2O_2]$ at RT	ECP at HT
	pH and conductivity at RT	$[H_2O_2]$ at HT
Processing	1) Estimating ECP	1) Estimating ECP at the location of interest
	2) Evaluating SCC propagation	2) Evaluating SCC propagation
Benefit	Sufficient database	1) High accuracy, 2) quick response
Disadvantage	Discrepancy between plant data and lab data	Under development (experience with long-term operation is required)

application to understand oxide film formation on stainless steel specimens as a result of the corrosion behaviour.

15.3   Experimental

15.3.1   High-temperature hydrogen peroxide water loop

A photograph of the high-temperature hydrogen peroxide loop is shown in Fig. 15.4. Details of the high-temperature, high-pressure H_2O_2 water loop and major experimental parameters have been shown in previous papers [5–7,11]. It was confirmed that more than 93% of H_2O_2 injected into the autoclave remained at the autoclave outlet and then test specimens were exposed to H_2O_2 with the lowest possible oxygen amount. The experimental conditions are listed in Table 15.2.

15.3.2   *In-situ* corrosive condition measurements

A schematic diagram of the ECP and FDCI measurement system is shown in Fig. 15.5. An external-type Ag/AgCl reference electrode was used to measure ECP.

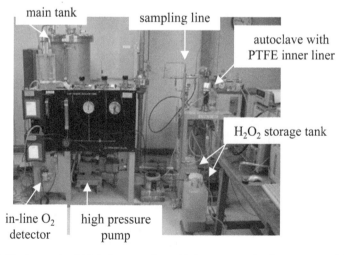

15.4 Photograph of high-temperature, high-pressure hydrogen peroxide water loop

Table 15.2   Major parameters for the experimental loop

Item	Parameter	Parameter range
Autoclave	Temperature	553 K
	Pressure	8.0 MPa
	Flow rate	2.8 ml s^{-1}
	Flow velocity	5.5 cm s^{-1}
	Conductivity	<0.2 μS cm^{-1}
	[O_2]	0–8000 ppb
	[H_2O_2]	0–1000 ppb
Feed water tank	Temperature	280–300 K
	Pressure	0.1 MPa
	Conductivity	<0.2 μS cm^{-1}
	[O_2]	0–8000 ppb
	[H_2O_2]	0 ppb

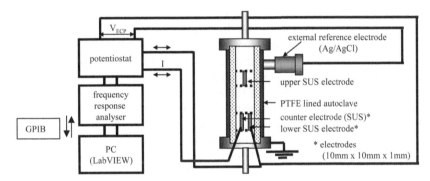

15.5 Schematic diagram of an autoclave for sensor array attached to the high-temperature H_2O_2 water loop

Upper and lower working electrodes of stainless steel were installed in the autoclave. ECP was measured with one of the parallel electrodes and [H_2O_2] was monitored by analysing FDCI for the parallel electrodes [6]. The coupled specimens were 10 × 10 × 1 mm thick. A sine wave alternating voltage was applied between the counter and lower work electrodes by changing the frequency from 0.1 mHz to 100 kHz to measure the complex impedance between the specimens. In order to minimise the potential input for the impedance measurement and perturbation of corrosion reaction on the specimens, the potential of one of the lower electrodes was kept as high as the corrosion potential against the reference electrode and minimum alternative voltage (10 mV) was applied to the other electrode for measurement [6].

15.3.3 Anodic polarisation current measurements

In the measurement of the polarisation properties in high purity water, a temporary reference electrode was applied to simplify the geometrical configuration around the specimens and the electrode and to fix the gap between the specimen and the electrode. As the temporary reference electrode, a Pt plate (10 × 10 × 1 mm) was used in the autoclave just in front of the specimen (3 mm away). Pt has often been applied as a reference electrode under high [H_2] conditions, but the Pt potential with lower [H_2]

varies with oxidant concentration. The potential of Pt was monitored and often calibrated against the external-type Ag/AgCl reference electrode to obtain the absolute value from the standard hydrogen electrode [15].

15.4   Experimental results

15.4.1   Relationship between ECP and FDCI

Measured ECP and Cole–Cole plots of FDCI of the specimens exposed to different [H_2O_2] are shown in Fig. 15.6 [6].

After 300 h exposure to 100 ppb H_2O_2, the concentration was reduced to 10, 5 and then 1 ppb. For each [H_2O_2], ECP was measured. The transient ECP is shown in Fig. 15.6a. Only a slight change in ECP was observed for the specimen when exposed to more than 5 ppb H_2O_2. When H_2O_2 was decreased to 1 ppb, the ECP decreased. Cole–Cole plots of measured complex impedance between a pair of stainless steel specimens exposed to H_2O_2 are shown in Fig. 15.6b [7]. The plots show the relationship between the real part of the complex impedance ($Re[Z(\omega)]$) and the negative of its imaginary part ($-Im[Z(\omega)]$). One half circle for low frequency from 1 mHz to 10 Hz, designated as the 'low frequency semicircle', was observed on the right side of the graph (the larger real part of impedance ($Re[Z(\omega)]$)), while the second for high frequency, but less than 10 kHz, designated as the 'high frequency semicircle', was observed in the smaller real part of the impedance. Only a slight change in ECP was observed for the specimens exposed to more than 5 ppb H_2O_2, while the radii of the Cole–Cole plots increased as [H_2O_2] decreased, which meant that the resistance of dissolution of oxide increased with unchanged resistance of oxide film. When [H_2O_2] decreased to 1 ppb, ECP decreased due to a drop in cathodic current at the specimen surface.

15.4.2   Effects of hydrogen peroxide and oxygen on ECP and FDCI

ECP and FDCI measured for the specimen exposed to a mixture of 100 ppb H_2O_2 and 200 ppb O_2 for 200 h are compared in Fig. 15.7 to a specimen exposed to just 100 ppb

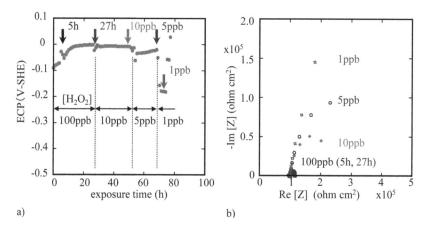

a)   b)

15.6 Electrochemical corrosion potential and complex impedance (changing [H_2O_2] after 300 h exposure to 100 ppb H_2O_2); a) ECP and b) complex impedance

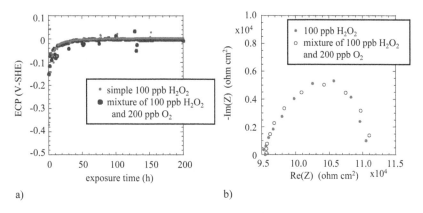

15.7 Electrochemical corrosion potential and complex impedance (100 ppb H_2O_2 and mixture of 100 ppb H_2O_2 and 200 ppb O_2); a) ECP and b) Cole–Cole plots of complex impedance

H_2O_2. No difference in ECPs and Cole–Cole plots could be observed. Neither dependencies of ECP nor saturation radii of the Cole–Cole plots on $[H_2O_2]$ were affected by co-existing O_2 at the same level of oxidant concentration [8].

ECP was measured first for the specimen exposed to 200 ppb O_2 for 200 h and then the corrosive conditions were changed from O_2 to H_2O_2 to measure ECP of the specimen exposed to 100 ppb H_2O_2 for 200 h. These measured ECPs are shown in Fig. 15.8a. When the corrosive conditions were changed to 100 ppb H_2O_2 after 200 h exposure to 200 ppb O_2, ECP increased to the plateau level (0 mV) in 30 h. Next, ECP was measured first for the specimen exposed to 100 ppb H_2O_2 for 200 h and then the corrosive conditions were changed from H_2O_2 to O_2 to measure ECP of the specimen exposed to 200 ppb O_2 for 200 h. When the corrosive conditions were changed to 200 ppb O_2 after 200 h exposure to 100 ppb H_2O_2, ECP decreased to the plateau level (–40 mV) in 10 h. The plateau level was higher than that for the specimen exposed to only 200 ppb O_2.

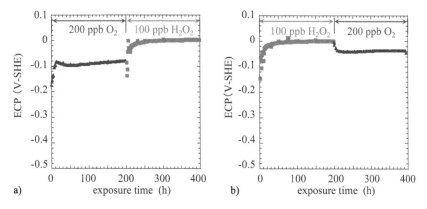

15.8 Measured electrochemical potential for combined O_2 and H_2O_2 exposures; a) Exposure to 200 ppb O_2 for 200 h after 200 h exposure to 100 ppb H_2O_2 and b) Exposure to 100 ppb H_2O_2 for 200 h after 200 h exposure to 200 ppb O_2

15.4.3   Modified ECP

From the polarisation measurements, it was seen that anodic current densities for H_2O_2 increased with $[H_2O_2]$ and at the same time, cathodic current densities also increased. Anodic current densities for H_2O_2 were also affected by oxide film on the specimen surface. ECP for O_2 exposure was determined by balancing the anodic and cathodic current densities as shown by the fine lines in Fig. 15.9 [15]. The corrosion current was also determined by the ECP. In the case of H_2O_2 exposure, ECP was determined by balancing the anodic and cathodic current densities (thick lines in Fig. 15.9), but the anodic current density at the ECP was not the same as the net corrosion current density. The net corrosion current density should be obtained by subtracting the H_2O_2 oxidation current density from the total current density. Then the potential which gave the same corrosion current density as the specimen exposed to O_2 could be designated as the modified ECP for discussion of crack growth rate for entire databases containing O_2 conditions and H_2O_2 conditions.

The crack growth rate of IGSCC is determined by the net corrosion current but not by the ECP [16]. However, most discussions of crack growth rates have been based on ECPs. The ECPs are different for specimens exposed to H_2O_2 and O_2. In order to discuss the crack growth rates with ECP, a modified ECP was proposed which gives the same corrosion current for both H_2O_2 and O_2 conditions and an empirical formula was derived to determine the modified ECP by calibrating the measured ECP with H_2O_2 oxidation current and electric resistance of oxide film [15]. As a result of application of the modified ECP, evaluation of crack growth rates showed a fairly good relationship with ECP for both O_2 and H_2O_2 (Fig. 15.10).

15.4.4   Equivalent circuit analysis

An equivalent circuit for the system in elevated temperature water containing sufficient amounts of oxidants is shown in Fig. 15.11 [17]. Each element of the complex

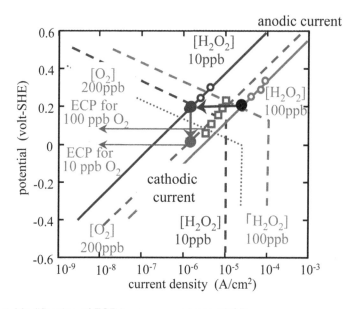

15.9 Modification of ECP by oxygen oriented ECP

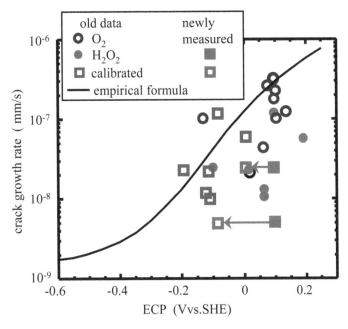

15.10 Evaluation of crack growth rate (modified ECP determined by measured ECP and corrosion current)

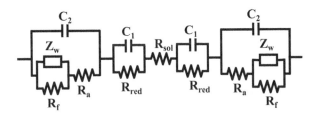

R_f: resistance of oxide film
R_a: resistance of oxide dissolution and H_2O_2 oxidation
R_{red}: resistance of redox reaction, H_2O/O_2 or H_2O/H_2O_2
R_{sol}: resistance of pure water between the specimen
Z_w: Warburg impedance $Z_w = Z (1-j)/(w)^{1/2}$
          Z : Warburg impedance coefficient
C_1, C_2: capacitances of oxide film/solution interface

15.11 Equivalent circuit model based on the corrosion model for film growth

impedance is also shown there. The sum of oxide dissolution resistance and H_2O_2 oxidation resistance was designated as the anodic resistance.

The radii of the low frequency semicircles were determined by the anodic resistance and oxide film resistance and for the specimens exposed to a higher $[H_2O_2]$, the effects of the anodic resistance on the radii decreased [11]. As a result of equivalent circuit analysis, the relationships between the radii of low frequency semicircles and $[H_2O_2]$ were obtained (Fig. 15.12). As $[H_2O_2]$ was lowered from 100 to 5 ppb, the radii of the low frequency semicircles increased continuously and the ECP remained at the same level down to 10 ppb and then decreased a little to 5 ppb. The increasing radii

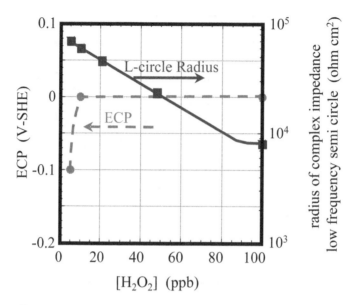

15.12 Relationships between [H₂O₂], ECP and FDCI after 200 h exposure

meant increasing anodic resistance. Both dependences of ECP and saturation of the radii on $[H_2O_2]$ were not affected by co-existing $[O_2]$ with the same level of oxidant concentration.

Unfortunately, it is difficult to separate the anodic resistance and oxide film resistance using only FDCI data. In order to separate them, oxide film resistance was determined by *in-situ* oxide film resistance [18].

15.5   *In-situ* sensor for hydrogen peroxide concentration

15.5.1   Sensor array

A schematic diagram of the sensor array to determine $[H_2O_2]$ in BWR primary cooling water is shown in Fig. 15.13. The sensor array consisted of an internal type

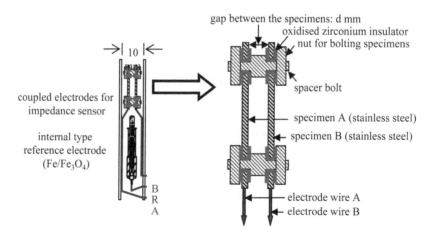

15.13 Schematic diagram of a sensor array

reference electrode and a coupled electrode for FDCI measurement. First ECP was measured between the reference electrode and specimen B of the couple, the secondary potential of specimen B was kept at the ECP level relative to the reference and then the alternative voltage was imposed on specimen A against B to measure current densities between A and B by changing frequency. With this procedure, surface change caused by electrically accelerated corrosion could be avoided.

The electric resistance of oxide film, R_f, has been measured as 1 kΩ cm^2 and then the anodic resistance, R_a, has been determined [19]. The low frequency semicircle was determined from the resistance of oxide film and resistance of oxide dissolution. The electric resistance of oxide film was much lower than that of oxide dissolution as can be seen in Fig. 15.14. The inverse of resistance of oxide dissolution is also shown, which is in proportion to [H$_2$O$_2$].

15.5.2   Reference electrode

In order to apply the sensor array in operating power plants or in pile loops, one of the key issues for selecting a suitable combination of sensors is their long-term reliability. Damage to the electrolyte of the reference electrode or the ceramics material used to seal the reference electrode often caused problems for the sensor itself. Cracks on the ceramic casing often led to leakage of liquid electrolyte or degradation of solid electrolyte. For instance, junction parts between the metal housing (Ni base alloy) and ceramics membrane (zirconia membrane) covering the Fe/Fe$_3$O$_4$ powder of Fe/Fe$_3$O$_4$ electrodes were often damaged by thermal cycling [20]. The most reliable sensor from the viewpoint of hardware has been the Pt sensor.

One disadvantage of the Pt sensor is that it is too sensitive to oxidant and hydrogen concentrations to determine the absolute ECP value. However, the purpose in applying the Pt sensor to the sensor array is not to determine the absolute ECP but to facilitate controlling the potential of the stainless steel specimens of the array at the

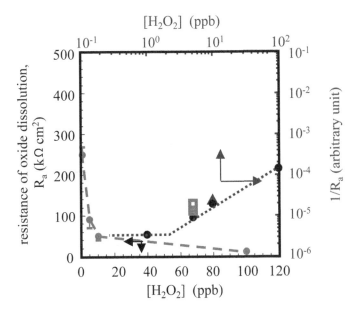

15.14 Relationship between [H$_2$O$_2$] and dissolution resistance

ECP level during measurement to moderate the effects of oxidant on stainless steel surface properties.

As a result of FDCI measurement, $[H_2O_2]$ was obtained from the relationship of resistance of oxide dissolution and $[H_2O_2]$ (Fig. 15.14), $[O_2]$ and $[H_2]$ were measured in the sampled water, and then the potential of the Pt sensor was estimated from the measured data. Then the potential of stainless steel specimens for measurement was calibrated by applying the estimated Pt potential.

Measured potentials of Pt plates exposed to H_2O_2 and O_2 are shown in Fig. 15.15 with those reported by Kim [21]. The potential of the Pt specimen exposed to O_2 was as high as the potential of the Pt specimen exposed to H_2O_2 at lower oxidant concentrations. The figure also shows the potential of Pt specimens exposed to O_2 or H_2O_2 calculated using the Evans Diagram [22]. The potential of specimens exposed to O_2 without co-existing H_2 monotonously increased with $[O_2]$, while that of specimens exposed to H_2O_2 reached the lower saturation level, where cathodic current densities due to the H_2O_2 reduction reaction and anodic current densities due to the H_2O_2 oxidation reaction balanced each other. In the conditions with co-existing H_2, the response of Pt electrodes might be much more complicated. However, the hardware of the Pt electrode is much more stable than the external-type Ag/AgCl reference electrode and the Pt electrode can be applied as a temporary reference electrode in high-temperature pure water when supported by suitable calibration.

An example of a sensor array with a Pt reference electrode is shown in Fig. 15.16.

15.5.3 Alternative data processing

Another disadvantage of applying the electrochemical sensor array for *in-situ* measurement of $[H_2O_2]$ in high-temperature water is in making a precise evaluation of Cole–Cole plots to determine oxidant concentration. Measured frequency dependencies of complex impedance for different $[H_2O_2]$ are shown in Fig. 15.17. As shown in

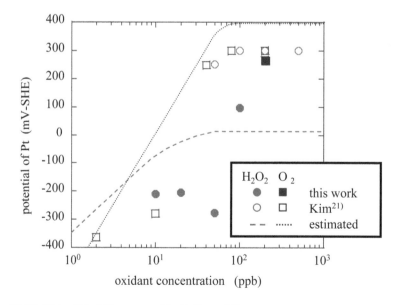

15.15 Potential of Pt exposed to H_2O_2 and O_2

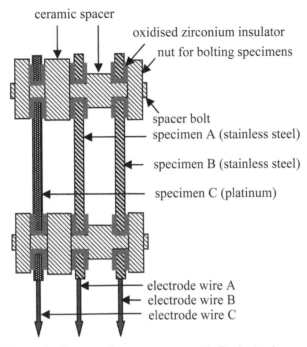

15.16 Schematic diagram of a sensor array with Pt electrode

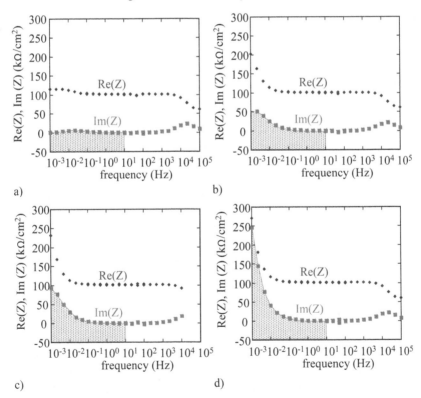

15.17 Frequency dependence of complex impedance; a) 100 ppb H₂O₂,
b) 10 ppb H₂O₂, c) 5 ppb H₂O₂ and d) 1 ppb H₂O₂

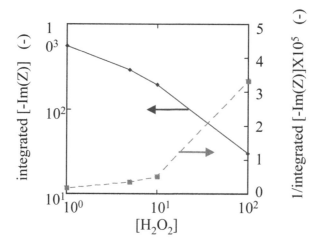

15.18 Relationships between [H₂O₂] and integrated imaginary part of complex impedance with less than 10 Hz frequency

Cole–Cole plots, oxidant concentration is obtained from the low frequency semicircles (frequency: <10 Hz). Instead of using Cole–Cole plots obtained with a frequency analyser, the integrated imaginary part of the frequency dependent complex impedance can also be applied to determine the oxidant concentration.

Relationships between [H₂O₂] and the integrated imaginary part of the complex impedance with less than 10 Hz frequency are shown in Fig. 15.18. Integration of the imaginary part of frequency dependent complex impedance is also applicable for oxidant concentration measurement.

15.6   Conclusions

The conclusions of this chapter and the studies presented herein are summarised as follows:

1. H_2O_2 concentrations could be determined by *in-situ* measurement of ECP and FDCI using sensor arrays. The ECP data could be used to evaluate the occurrence and propagation of IGSCC of reactor internal components.
2. The sensor array, consisting of ECP and FDCI sensors for corrosive condition monitoring, was proposed for application in operating power plants.
3. The Pt reference electrode was applicable as a stable reference electrode for the sensor array.

Acknowledgements

The study has been supported by the Japan Society for the Promotion of Science (JSPS) [A Grant in Aid for Scientific Research: Subject No. 16360467 (2004–2006)]. As a result of transfer of the chief researcher of the programme from Tohoku University to the Japan Atomic Energy Agency (JAEA) in April 2005, the experimental facilities were moved. The authors express their sincere thanks to the JSPS and the JAEA for supporting the experiments.

References

1. S. Uchida, E. Ibe, K. Nakata, M. Fuse, K. Ohsumi and Y. Takashima, *Nucl. Technol.*, 110 (1995), 250–257.
2. Y. Wada, S. Uchida, M. Nakamura and K. Akamine, *J. Nucl. Sci. Technol.*, 36 (1999), 169–178.
3. Y. J. Kim, L. W. Niedrach and C. C. Lin, 'Development of ECP models for BWR application', in *Proc. of 7th Int. Symp. Environmental Degradation of Materials in Nuclear Power Systems – Water Reactors*, Breckenridge, CO, 699–706. National Association of Corrosion Engineers, 1995.
4. R. J. Row, M. E. Indig and C. C. Lin, 'Suppression of radiolytic oxygen produced in a BWR by feedwater hydrogen addition', in *Proc. Int. Conf. Water Chemistry of Nuclear Reactor Systems, Water Chemistry 3*, 23. British Nuclear Energy Society, 1984.
5. T. Satoh, S. Uchida, J. Sugama, N. Yamashiro, T. Hirose, Y. Morishima, Y. Satoh and K. Iinuma, *J. Nucl. Sci. Technol.*, 41 (2004), 610–618.
6. J. Sugama, S. Uchida, N. Yamashiro, Y. Morishima, T. Hirose, T. Miyazawa, T. Satoh, Y. Satoh, K. Iinuma, Y. Wada and M. Tachibana, *J. Nucl. Sci. Technol.*, 41 (2004), 880–889.
7. T. Miyazawa, S. Uchida, T. Satoh, Y. Morishima, T. Hirose, Y. Satoh, K. Iinuma, Y. Wada, H. Hosokawa and N. Usui, *J. Nucl. Sci. Technol.*, 42 (2005), 233–241.
8. V. A. Marichev, *Corros. Sci.*, 38 (1996), 531–558.
9. H. Takiguchi, S. Sekiguchi, A. Abe, K. Akamine, M. Sakai, Y. Wada and S. Uchida, *J. Nucl. Sci. Technol.*, 36 (1999), 179–188.
10. S. Uchida, K. Ishigure, H. Takamatsu, H. Takiguchi, H. Nakagami and M. Matsui, 'Chemistry data acquisition, processing, evaluation and diagnosis systems for nuclear power reactors', in *Proc. 14th Int. Conf. Properties of Water and Steam* (Kyoto, Japan), 551–556. International Association for Properties of Water and Steam, 2004.
11. S. Uchida, T. Satoh, J. Sugama, N. Yamashiro, Y. Morishima, T. Hirose, T. Miyazawa, Y. Satoh, K. Iinuma, Y. Wada and M. Tachibana, *J. Nucl. Sci. Technol.*, 42 (2005), 66–74.
12. E. Ibe and S. Uchida, *Nucl. Sci. Eng.*, 89 (1985), 330–350.
13. T. Satoh, S. Uchida, Y. Satoh and T. Tsukada, 'Effect of hydrogen peroxide on corrosion of stainless steel in high temperature water', in *Proc. 15th Int. Conf. Nuclear Engineers (ICONE15-10607)*, 22–26. Japan Society of Mechanical Engineers, Nagoya, Japan, 2007.
14. N. Yamashiro, S. Uchida, Y. Satoh, Y. Morishima, H. Yokoyama, J. Sugama and R. Yamada, *J. Nucl. Sci. Technol.*, 41 (2004), 890–897.
15. S. Uchida, Y. Morishima, T. Hirose, T. Miyazawa, T. Satoh, Y. Satoh and Y. Wada, *J. Nucl. Sci. Technol.*, 44 (2007), 758–766.
16. Y. Wada, A. Watanabe, M. Tachibana, N. Uetake, S. Uchida and K. Ishigure, *J. Nucl. Sci. Technol.*, 37 (2000), 901–912.
17. K. Mabuchi, M. Sakai and N. Ohnaka, *Zairyo-to-Kankyo*, 43 (1994), 2–10 (in Japanese).
18. S. Uchida, T. Satoh, N. Kakinuma, T. Miyazawa, Y. Satoh and K. Mäkelä, *ECS Trans.*, 2(25) (2007), 39–50.
19. T. Satoh, S. Uchida, T. Tsukada, Y. Satoh and K. Mäkelä, 'The effect of oxide film on electrochemical corrosion potential of stainless steel in high temperature water', in *Proc. 13th Int. Symp. on Environmental Degradation of Materials in Nuclear Power Systems – Water Reactors* (Whistler, BC, Canada, 19–23 August 2007). National Association of Corrosion Engineers, CNS (CD-ROM).
20. T. Tsukada, Y. Miwa and H. Ugachi, 'In core ECP sensor designed for the IASCC experiments at JMTR', in *Proc. Int. Conf. Water Chemistry of Nuclear Reactor Systems*, P2/25 (San Francisco, CA, USA, 11–14 October 2004). Electric Power Research Institute (CD-ROM).

21. Y. J. Kim, 'Electrochemical interactions of hydrogen, oxygen and hydrogen peroxide on metal surfaces in high temperature, high purity water', in *Proc. 8th Int. Symp. on Environmental Degradation of Materials in Nuclear Power Systems – Water Reactors* (America Island, FL, 10–14 August 1997), 641–650. National Association of Corrosion Engineers, ANS.

22. S. Uchida, T. Satoh, T. Tsukada, Y. Satoh and Y. Wada,'Comparison of the effects of hydrogen peroxide and oxygen on oxide films on stainless steel and behaviors in high temperature water', in *Proc. 13th Int. Symp. on Environmental Degradation of Materials in Nuclear Power Systems – Water Reactors* (Whistler, BC, Canada, 19–23 August 2007). National Association of Corrosion Engineers, CNS (CD-ROM).

Index

Note: Page references in *italics* refer to illustrations

ACM Instruments 120
acoustic emission (AE) tech-niques
        7, 120−1,135−41
Alloy 600 81−2, 89, 92−4
all-volatile treatment (AVT) of feed
        water 221−2
alternating current potential drop
        (ACPD) technique 6
Anderson, D. 21, 25
ANDRA (French Agency for Nuclear
        Waste Man-agement) 143
Anita, T. 87
anodic polarisation current
        measurements 243−6
anodic reactions 125−6, 132, 141
anodic resistance 247−8
anodic Tafel parameters 10−12, 19
austenitic steels 46−7, 63−4, 97, 179
autoclave equipment 145, 212−17,
        222−3, 228−32, 237, 243

Bogaerts, W.F. 22, 25
boiling water reactor (BWR)
        technology 8, 32, 46−7, 61, 96,
        63−4, 79, 159−68, 171−5, 182,
        188−92, 194−5, 208−10, 212,
        214, 235, 239−42, 248
Bosch, R.W. 22

CANDU power stations 222, 226, 234
carbon steel components 143−56, 194,
        234−7
cathodic current densities 246
cathodic reactions 125−6, 141
cathodic reduction of water 155
cathodic Tafel parameters 10−12, 19
cationic vacancies 108
cladding of fuel rods 189−92
clay barriers for nuclear waste sites
        143−56

closed cooling water sys-tems 205−6
CLT test 56−9
Cole-Cole plots 155, 165−6, 244−5,
        250−2
compact tension (C(T)) specimens
        173−4
complex capacitance spectra 146−7,
        156
constant phase elements (CPEs) 21,
        111−12, 147, 156
contact resistance 7, 96
controlled distance electro-chemistry
        (CDE) 96
cooling systems of nuclear power
        plants 158−68; *see also* closed
        cooling water systems
CORD process 199−203
corrosion coupons 199−201
corrosion potential 7−10; *see also*
        electrochemical corrosion
        potential
crack growth rates (CGRs) 8−9, 197−9,
        247

decontamination processes 194,
        199−203
Dévay, J. 22
dielectric response 104, 115
diffusion impedance 113−15
direct current potential drop (DCPD)
        technique 6, 47−61, 94, 173, 189,
        206−8
doping density 106
double layer capacitance 111

Einstein-Stokes equation 114−15
Electric Power Research Institute
        (EPRI) 222
electric resistance (ER) technique 6−7
Electricité de France 222

electrochemical corrosion potential
(ECP) 47–9, 161, 165–8, 181–7,
194–7, 203–6, 210–12, 216,
218–19, 224, 237, 239–52
electrochemical current noise
(ECN) 32, 49, 54, 56, 81, 87–90
electrochemical frequency modulation
(EFM) 5–6, 22, 25
electrochemical impedance spectroscopy (EIS) 5, 13–16, 25, 96, 99,
146, 155, 191
electrochemical methods of corrosion
monitoring 5–28; accuracy of
3–4; hybrid 120
electrochemical noise (EN) 6, 32–43,
46–61, 64–8, 73–5, 79, 81, 92,
94, 96, 99, 120, 123, 126, 140–1,
199–202; during stress corrosion
cracking 37–41; generation of
39–44
'electrochemical nose' tech-nique 6, 25
electrochemical potential noise (EPN)
32, 49–61, 81, 87–91
energy-dispersive X-ray spectroscopy
(EDS) 150–4
equivalent circuit analysis 246–8
Evans, U.R. 222
Evans diagram 250

Faraday's law 110
Ferreira, P. 25
field signature method (FSM) 6–7
flow assisted corrosion (FAC) 222
fractographic analysis 69–72
Framatome ANP GmbH 199
frequency dependent com-plex
impedance (FDCI) sensors
165–6, 239–45, 248–52
Fukuda, T. 87

gamma spectrometry 173
Gouy-Chapman layer 112–13
Graham, M.J. 154

Halden Reactor Project 171–92
harmonic analysis (HA) 4–6, 18–22, 25
harmonic impedance spectroscopy
(HIS) 20–1
Helmholtz layer 112

Hino, Y. 234
Hwang, S.S. 64
hydrazine concentrations 167, 203–5,
211–12, 221–4, 228, 231–4, 238
hydrogen-assisted cracking (HAC) 34
hydrogen-induced brittle fracture 132,
134, 141
hydrogen-induced cracking (HIC) 33–4
hydrogen peroxide, use of 239, 242–5,
248–52
hydrogen water chemistry (HWC) 8,
166, 173, 175, 194–9, 211–12,
240–1

ideal corrosion monitoring method,
capabilities of 1
impedance measurements 99–115; see
also dif-fusion impedance
Inconel 690 234, 237
inductively coupled plasma mass
spectrometry 173
in-pipe corrosion monitor-ing 212–16
intergranular attack (IGA) 128–30, 221
intergranular stress corrosion cracking
(IGSCC) 63–4, 69, 77–9, 194,
211, 239–40, 246, 252
intermodulation distortion technique
see electro-chemical frequency
modulation
International Atomic Energy Agency
(IAEA) 163
ion chromatography 167
irradiation-assisted stress corrosion
cracking 173

Kendig, M. 21, 25
Kihir, H. 21
Kim, Y.J. 250
Kuş, E. 23

Lawson, K. 19
light water reactor (LWR) technology
158, 171, 173
linear polarisation resist-ance (LPR)
technique 5, 7, 11, 25, 205
linear voltage differential transformers
(LVDTs) 176, 180–3
localised corrosion 5, 16, 124–6, 221

localised corrosion monit-oring (LCM)
    120–2, 130–3, 140–1
Lowson, R.T. 222
Lukac, C. 153–4

Macdonald, D.D. 64, 235
magnesium partially stab-ilised zirconia
    (MPSZ) 186
Mansfeld, F. 23
Mészáros, L. 19, 22
Mizuno, T. 87
Montemor, M.F. 97
Mössbauer transmission spectroscopy
    226, 228, 232, 237
Mott-Schottky plots 107–8
Multeq code 234–7

Nakamura, T. 234
Newman, R. 69
noise resistance 99–100
normal water chemistry (NWC) 166,
    173–6, 195–9, 241
nuclear waste disposal 143
Nyquist diagrams 15–16, 20

ohmic resistance 11, 15
Organization for Economic Coopera-
    tion and Develop-ment (OECD)
    Nuclear Energy Agency 171
oxide layer, impedance of 102–8

palladium (Pd), use of 182–8
passivation 148
PbSCC propagation 85–94
pipe-thinning monitors 205–8
pitting corrosion 21, 25
plastic deformation 137, 141
point defect model 154–6
Poisson distribution 92–3
polarisation and polarisation resistance
    10–15, 21, 99–100, 108–15
Pourbaix diagrams 7–8, 234
power density spectrum (PDS) 73–7
pressurised heavy water reactor
    (PHWR) tech-nology 171–2,
    179, 188, 192
pressurised water reactor (PWR)
    technology 9, 32, 46, 64, 81, 96,
    160–2, 167–8, 194, 203–13, 216,
    219, 221–6, 234, 237

probes for corrosion monit-oring 1–3
programmable logic control (PWC) 172
PSD plots 89, 91
Pyun, S.I. 93

radiation induced segreg-ation (RIS) 63
redox conditions and redox potentials
    211–12, 222, 226, 232–7
reference electrodes 8–9, 144–5, 181,
    184–8, 192, 218, 249–50
repassivation processes 34, 39–44,
    78–9, 141
risk-benefit balance 168
Rosborg, B. 20
rupture frequency 41

Sawicki, J.A. 234
Sawochka, S.G. 234
Scantlebury, J.D. 19
scratch tests 34
SEM observations 150–4
semiconductors 104–5
Shibata, T. 92
simultaneous use of corros-ion
    detection techniques 141
slow strain rate tests (SSRTs) 64–7
Solomon, H.D. 69
stainless steel components 46–7, 64,
    104–5, 120–2, 140, 179, 194–7,
    224–3, 237, 239, 243, 250
standard hydrogen elec-trode (SHE)
    181
steam generators 96
Stern-Geary equation 3–4, 11–12, 15
Stewart, J. 69, 74
stress control cracking (SCC) 32–43,
    46–61, 63, 81, 120–41, 173–5,
    194, 197–9, 221

Tafel extrapolation 5, 10–12, 19–20
Taiwanese power plants 194–210
thiosulphate, use of 132
time evolution of transfer conductance
    148–50
transfer functions 15–16; see also time
    evolution of transfer conductance
transgranular stress corros-ion cracking
    (TGSCC) 64, 69, 74, 77–9
transients, analysis of 87, 122–35

ultrasonic testing (UT) 7
uniform corrosion 5, 93–4, 124
Urquidi-Macdonald, M. 235

Warburg coefficients 113–15
water chemistry 158–61, 171–2,
    211–12, 219, 239; data sources
    for 162; use of high-temperature
    sensors 162–8; *see also* hydrogen
    water chemistry; normal water
    chemistry
wedge-opening-loaded (WOL)
    specimens 199

Weibull distribution func-tion 89, 94
Wells, D.B. 69, 73, 77

X-ray diffraction analysis 145, 150–2
X-ray fluoroscence spec-troscopy
    (XRF) 173, 229–32

yttrium partially stabilised zirconia
    (YPSZ) 186
yttrium stabilised zirconia (YSZ) 9

zero resistance ammeters (ZRAs) 126
Ziracloy fuel cladding 191–2
zirconium oxide 104